REFORESTING THE EARTH

SOCIETY AND THE ENVIRONMENT

SOCIETY AND THE ENVIRONMENT

The impact of humans on the natural environment is one of the most pressing issues of the twenty-first century. Key topics of concern include mounting natural resource pressures, accelerating environmental degradation, and the rising frequency and intensity of disasters. Governmental and nongovernmental actors have responded to these challenges through increasing environmental action and advocacy, expanding the scope of environmental policy and governance, and encouraging the development of the so-called "green economy." Society and the Environment encompasses a range of social science research, aiming to unify perspectives and advance scholarship. Books in the series focus on cutting-edge global issues at the nexus of society and the environment.

SERIES EDITORS

Dana R. Fisher
Lori Peek
Evan Schofer

Super Polluters: Tackling the World's Largest Sources of Climate-Disrupting Emissions, Don Grant, Andrew Jorgenson, and Wesley Longhofer
Underwater: Loss, Flood Insurance, and the Moral Economy of Climate Change in the United States, Rebecca Elliott

REFORESTING THE EARTH

· · · · · · · · · · · · · · · · · · ·

The Human Drivers of Forest Conservation,
Restoration, and Expansion

THOMAS K. RUDEL

COLUMBIA UNIVERSITY PRESS

New York

Columbia University Press
Publishers Since 1893
New York   Chichester, West Sussex
cup.columbia.edu

Library of Congress Cataloging-in-Publication Data
Names: Rudel, Thomas K., author.
Title: Reforesting the earth : the human drivers of forest
conservation, restoration, and expansion / Thomas K. Rudel.
Description: New York : Columbia University Press, [2023] | Series: Society
and the environment | Includes bibliographical references and index.
Identifiers: LCCN 2023004282 (print) | LCCN 2023004283 (ebook) |
ISBN 9780231210683 (hardback) | ISBN 9780231210690 (trade paperback) |
ISBN 9780231558549 (ebook)
Subjects: LCSH: Reforestation—Climatic factors. | Reforestation. | Afforestation.
Classification: LCC SD409 .R84 2023 (print) | LCC SD409 (ebook) |
DDC 634.9/56—dc23/eng/20230407
LC record available at https://lccn.loc.gov/2023004282
LC ebook record available at https://lccn.loc.gov/2023004283

Printed in the United States of America

Cover design: Noah Arlow
Cover image: Shutterstock

In memory of Maria Delores Quesada

We are tenants of the Almighty
Entrusted with a portion of His earth
To dress and keep
And pass on to the next generation
When evening comes and we must fall asleep.

—LOUISIANA DUNN THOMAS

Contents

Preface   xi

Acknowledgments   xv

List of Abbreviations   xvii

1   Forests: A Natural Climate Solution   1

2   Theory: Societal Transformations, Corporatism, and Forest Gains   12

3   Forest Losses, the Conservation Movement, and Protected Areas   42

4   Rural–Urban Migration, Land Abandonment, and the Spread of Secondary Forests   68

5   Planted Forests: Concessions, Plantations, and the Strength of States   97

6   Agroforests I: The Spread of Silvopastures   128

7   Agroforests II: Restoring Agroforests in the Humid Tropics   149

8 Resurgent Forests: A Qualitative Comparative Analysis   175

9 A Global Forest Transition?   198

Glossary   217
Notes   219
Bibliography   247
Index   269

Preface

Reforesting the Earth has reached the public during a period of paramount uncertainty about climate change. Since the early 2000s, climate change has hit human communities with unexpected force at the same time that political conflict over the move away from climate-changing fossil fuels has intensified. In this context, interest in natural solutions to climate change has accelerated. These solutions promise to reduce the scale of the changes in climate by accelerating the absorption of carbon by forests, grasslands, and wetlands. The increased carbon sequestration would stem largely from increases in carbon-rich forests. For landscapes to become carbon sinks, a reversal of long-standing declines in the extent of forests must occur. In addition to gains from sequestered carbon, recovering forests would in many instances provide additional, incalculable benefits by bolstering the preservation of biodiversity in landscapes. These benefits from resurgent forests have led numerous observers to ask how, exactly, societies spread forests across landscapes.

Land-change scientists from the natural and the social sciences have attempted to answer this question in a long series of articles in academic journals. With the accounts of forest resurgence written by scientists from diverse disciplines and published in a wide disciplinary range of journals, knowledge about how societies have reforested their landscapes has been slow to accumulate. This book seeks to remedy this situation by identifying patterns that have run across the diverse cases of reforestation discussed in the literature. With luck, knowledge about these patterns will provide activists and policymakers with guidance in their efforts to implement natural climate solutions to climate change.

The analytic work reported here began with case studies of forest conservation, recovery, and expansion carried out by a diverse set of researchers. Anthony Anderson, Jeffrey Bentley, Amy Daniels, James Goldthwait, Robert Heilmayr, Patrick Meyfroidt, Chris Reij, Ximena Rueda, and John Zinda carried out the detailed case studies that have provided the empirical foundations for the following analysis. The analytic work reported here stands on their shoulders.

Researchers working within two intellectual frameworks, land-change science and forest-transition theory, have together posed the questions that have guided the analyses reported here. Land-change science, built through studies of newly available, high-resolution satellite imagery, has made the quantitative study of forest cover change possible at a wide range of scales. Research with satellite images, supplemented by field research, has made it possible to map the geographical contours of forest resurgence wherever it has occurred. Works within this intellectual tradition by Eric Lambin and Patrick Meyfroidt have been particularly useful in suggesting the dynamics that have driven forest resurgence.

Analyses of forest transitions, long-term shifts in forest cover trends from deforestation to reforestation, have provided a second set of questions that have shaped the analyses reported here. Initiated by Alexander Mather in the early 1990s, research on forest transitions always sets forest recoveries in an historical context and explains why forests reappeared in particular locales at given points in time. Mather's work reminds me of the never-ending need to emphasize the historical dimensions of forest recoveries. For discernible reasons, they begin, grow, and decline over time. I try to identify the drivers of these historical changes in the following pages.

My own personal history in and around forests has influenced the analyses reported here in ways that should be acknowledged. I grew up in the New England suburbs. The highlights of my adolescence were fishing trips with my dad that took us into the forests of northern New England and French-speaking Quebec. I came to appreciate the beauty and the biodiversity of these northern forests. Closer to home, expanding metropolitan areas replaced forests with houses. As I watched these changes in land-scapes, the contingent nature of forests became clear. They could and did disappear.

After college, I joined the Peace Corps and found myself working in a state-supported, new land-settlement scheme on the fringes of the Ama-zon basin in southeastern Ecuador. This program delivered middle-sized plots of forested land to landless peasants from the Andes. The colonists mowed down the rainforest on their newly acquired lands, planted pas-tures, and began raising cattle. The resident Indigenous people, the Shuar, lost hunting grounds to the colonists, even as they acquired tracts of land around their homesteads that they could call their own.

Deforestation brought all manner of changes to these places. Rainfall patterns shifted, and landscapes dried out. A single species of pasture grasses covered soils where hundreds of different tree species had grown. The decline in biodiversity with the destroyed forests was most palpable in the early mornings. Before the land clearing began, all manner of screeches and howls from birds, cats, and wild pigs occurred in the predawn darkness. A year later, with large swathes of the forest removed, the only early morn-ing sounds came from the battery-driven radios of the colonists. The few remaining forests had gone silent.

My firsthand experiences with the destruction of tropical forests fueled my later research endeavors, both in trying to understand the drivers of tropical deforestation and the forces behind forest recoveries. A desire to contribute in some small but meaningful way to the alleviation of the global climate crisis also pushed me to do the synthetic work, reported here, on the circumstances that have shaped successful human efforts to con-serve, restore, and expand forests. I hope it will have an enabling effect, making clear the strategies and circumstances that have contributed to for-est conservation, recovery, and expansion across a range of places. These examples could encourage activists to build similar sorts of organizational alliances in other locales where success in efforts to conserve, restore, or expand forests seems conceivable.

Acknowledgments

A loosely structured community of scholars and scientists have helped to both sustain and discipline me as I did the work reported here. Steve Brechin, through his early insistence that organizations matter in the construction of sustainable societies, set me on the analytic path reported here. Ruth Chazdon, through her espousal of an ecological point of view, made me more aware of the ecological dimensions in forest recoveries. Sean Sloan, through his engagement with questions about the dynamics of reforestation, made me more aware of the anomalous patterns of reforestation in both the Americas and Southeast Asia. Brad Walters, Julia Flagg, and one of the reviewers for Columbia University Press made detailed edits of the text that made it much more readable. Mike Gildesgame, throughout a two-year period of writing, brought breaking stories about deforestation and reforestation to my attention. Diane Bates, Chun Fu, Heather Zichal, and Marla Perez-Lugo, while students at Rutgers University, provided valuable help in assembling and analyzing data on forest losses and recoveries. Megan McGroddy, Amy Lerner, Diana Burbano, Maria Delores Quesada, and Gerardo Caivinagua provided indispensable assistance in the collecting and analyzing the data on the emergence of silvopastoral landscapes in the upper reaches of the Amazon basin. Mike Siegel made the map in chapter 8. Two grants from the National Science Foundation (SBR9618371 and CNH10009499) and funds from the United States Fulbright Commission for Ecuador provided crucial financial support for the investigations of park creation and silvopastoral expansion in the Ecuadorian Amazon.

I owe a debt of gratitude to the people at Columbia University Press for transforming a dry and sometimes wordy manuscript into a book. Miranda Martin shepherded it through a multitude of steps to publication. Dana Fisher, Lori Peek, and Evan Schofer, the editors of the Society and the Environment series at Columbia, provided support for the manuscript at a critical juncture in the publication process. Three anonymous reviewers for CUP provided very useful feedback on an early draft of the manuscript that guided my efforts to revise the manuscript for publication. Brian C. Smith clarified a large number of stylistic issues that arose in preparing the manuscript for publication.

I researched and wrote this manuscript during the coronavirus pandemic of 2020–2022. The isolation from interactions with other people may have accelerated the work on *Reforesting the Earth*, but it weighed on us all. My wife, Susan, and our son, Dan, endured the isolation like me, but I made it worse for them by retreating to my computer. I hope this book will provide some recompense for the lost hours with family and friends.

Finally, let me add a few words about the person to whom this book is dedicated. Maria Delores Quesada worked on and managed several research projects with me in the Ecuadorian Amazon between 1995 and 2015. On innumerable occasions, her observations about changing tropical landscapes had a clarifying effect for the rest of us. The mother of five children, she had a lifelong knack for organizing the women in the villages where we did research on landscape changes. She figured out ways to make research funds meet the goals of the grant at the same time that the funds delivered benefits to the communities. For example, to reward respondents for participating in one study, she came up with idea of giving refrigerators to the daycare centers in the villages where we did the research. The refrigerators were much needed and much appreciated in this tropical setting, in which milk spoils easily. Unfortunately, Ms. Quesada died in a motorcycle accident in 2020. The dedication is meant to honor her memory.

Abbreviations

C	carbon
CAP	common agricultural policy
CO_2	carbon dioxide
COP	conference of parties
EFA	ecological focus area
EU	European Union
FSC	Forest Stewardship Council
GBM	Green Belt Movement
GGW	Great Green Wall
IMF	International Monetary Fund
IPCC	Intergovernmental Panel on Climate Change
IUCN	International Union for the Conservation of Nature
NGO	nongovernmental organization
PES	payment for environmental services
REDD+	reduced emissions from deforestation and degradation
RFFP	Returning Farmland to Forest Program
SSA	sub-Saharan Africa
SCPP	Sustainable Cocoa Production Program

REFORESTING THE EARTH

1

• • • •

Forests

A NATURAL CLIMATE SOLUTION

Amid the growing impacts of climate change on human communities and the persistent inability of industrial societies to curb the burning of fossil fuels, scientists have increasingly looked at an expansion in the extent of the earth's forests as a natural solution to the climate crisis.[1] Expanding forests would sequester carbon above and below ground, in effect creating a negative emissions scenario in which growing trees would suck carbon out of the atmosphere.[2] The magnitude of this carbon sequestering effect could be considerable. One recent review of carbon sequestration by trees concluded that it could account for as much as 30 percent of the emissions declines necessary to keep global warming below 2°C.[3] In this context, it becomes important to answer questions about the circumstances in which gains in forest cover occur. By clarifying the determinants of forest recovery, this book points to a path that activists in the climate change movement might want to pursue in promoting natural climate solutions to the crises posed by global warming.

Forests in Flux

The centrality of forests in arguments about mitigating the effects of climate change presumes that forests, with human inducements, could expand in a warming world. By some measures, forests have already increased in extent. Total forest area on Earth grew by 7.1 percent between 1982 and 2016, as increases in forest cover in the Global North outpaced declines in the Global South.[4] In addition, scientists have maintained for three decades that the growing volume of greenhouse gases in the atmosphere could intensify photosynthesis and, in so doing, spur growth in leaf areas and carbon sequestration. Leaf areas have increased. China showed particularly large increases in leaf areas between 2000 and 2017. The coincidence of the leaf area increases and the Chinese government's ambitious promotion of small-scale tree plantations suggests that government policy may have played a role in the overall increase in leaf areas.[5]

There are, however, good reasons to doubt the climate stabilizing effects of forest cover trends. With additional warming, vapor pressure deficits have increased, and the associated dryness has begun to limit tree growth, reducing the stature of young trees and diminishing but not eliminating the ability of forested landscapes to act as sinks for atmospheric carbon.[6] Precipitation nourishes forests, so changes in amounts of rainfall have important implications for the carbon sequestering capacity of forests. Expected and observed rainfall differed substantially across regions during the late twentieth century. Overall, landscapes dried out, beginning in the 1950s.[7] Reductions in soil moisture through increased evapotranspiration at higher temperatures accounted for the drying out of landscapes in Africa, in the Mediterranean basin, in East Asia, and in South Asia.[8] Under these conditions, droughts and heat stressors on trees grew in frequency and in magnitude. As a consequence, tree mortality increased. These die-offs, if large in extent, could have converted forests from carbon sinks into carbon sources.[9]

Climate change threatens forests through fires as well as droughts. Drought exacerbates fires by making the dying trees more combustible, so droughts have often prepared the ground for massive fires that, among other things, emitted large stocks of carbon into the atmosphere when the woody biomass in forests burned. This link between forest fires, seasonally dry periods, and droughts has been obvious in Southeast Asia since the 1980s.[10] If the forest fires are extensive enough, they could destroy enough forests to erase any recent gains in overall forest area. This sequence of events seems to have occurred most recently in Portugal. It experienced a substantial forest transition during the twentieth century when deforestation ceased and reforestation began. Forest cover increased from 7 percent to 40 percent of the land area between 1875 and 2000.[11] Since then, wildfires have deforested extensive areas of southern Portugal and killed more than a hundred people.[12] Beginning in the 2010s, warming trends have also precipitated sharp increases in fires in the extensive peatlands of Russia's boreal forests.[13]

The links between climate change, droughts, fires, and forest transitions are not uniform. Grasslands and savannas have seen reductions in the extent of fires since the 1990s as intensified agriculture in zones such as the Sahel has discouraged the burning of croplands and pastures.[14] At the same time, the connections between global warming and lengthened fire seasons in Russia's boreal forests and America's Pacific Coast forests seem indisputable. The line of causation running from climate change to droughts, forest fires, and diminished prospects for forest restoration would seem then to have a regional basis, with climate change–induced fires destroying forests in some regions while collective action spurs forest growth and regrowth in other regions. This variety in recent forest outcomes suggests a plasticity in the conditions of forests that humans could shape to their own ends.

Natural Climate Solutions

Faced with adverse environmental trends, observers have rightly begun to search for ways to supplement necessary cuts in fossil fuel emissions with growth and regrowth in forested landscapes which would sequester carbon from the atmosphere. Landscape changes, as policy measures to contain climate change, are both inexpensive and effective. As a means for reducing greenhouse gas emissions, the maintenance of existing stands of forests costs only 20 percent as much as it would cost to reduce an equivalent amount of greenhouse gas emissions through reengineered industrial activities.[15] One high profile simulation of a Natural Climate Solution would restore about 15 percent of the current agricultural land base to forests. In so doing, this kind of policy initiative might counter up to 30 percent of the additional CO_2 emissions since the Industrial Revolution.[16] Taken together, the top twenty natural climate solutions could provide between 12 percent and 37 percent of the CO_2 reductions required to make it 66 percent probable that global warming would be limited to less than a 2°C rise.[17]

A glance at carbon removal rates for forests suggests that the most rapid rates of carbon sequestration occur in humid places. The rainfall spurs the growth of vegetation, and, as the plant grow, they sequester carbon. Rainforests have the world's largest above ground stocks of carbon. In Southeast Asia, the stocks of carbon in primary rainforests range from 254 to 647 MgC ha^{-1}.[18] Rates of carbon removal from the atmosphere show the expected pattern with the highest rates of removal occurring in primary tropical rainforests.[19] One exception to this overall pattern involves planted landscapes in which eucalyptus trees (*Eucalyptus globulus*), with exceptionally high rates of CO_2 removal, predominate.[20] The large stocks of carbon in tropical forests, coupled with the large scale of the ongoing forest destruction in the Global South, makes it the primary site for contemporary struggles to implement natural climate solutions to climate change.

Where human societies fall in the range between the 12 percent and the 37 percent abatement scenarios from forest regrowth depends on the willingness of societies to pursue negative emissions strategies. Societies would achieve only a 12 percent reduction in emissions if they decided to limit their expenditures on carbon sequestration to scenarios where abatement costs less than $10 per megagram of CO_2 abatement. If societies were to commit to spending up to $100 per megagram of CO_2 abatement, then the magnitude of the decline in greenhouse gases from negative emissions could be as high as 37 percent.[21] In sum, natural climate solutions have the potential to play a significant, worldwide role in extracting humans from the predicaments created by climate change. This prospect underlines the importance of questions about the ways in which natural climate solutions actually happen. This book addresses and tries to answer these questions.

Top-Down Approaches to Natural Climate Solutions

The recent history of high profile, top-down approaches to natural climate solutions suggests some of the challenges associated with these solutions. As the gravity of the climate change crisis has become more apparent, activists have increased their efforts to implement natural climate solutions. Officials from the United Nations and delegates from nations had laid the groundwork for collective action through the creation of two high-profile initiatives in 1992 when they met at the Earth Summit in Rio de Janeiro. The delegates drafted the Framework Convention on Climate Change and the Convention on Biological Diversity. These treaties, subject to approval by individual nations, authorized collective actions to forestall climate change and preserve biodiversity, respectively. To achieve these ends, delegates from the participating nations have met periodically ever since 1992 to reaffirm their commitments to the goals of the conventions and chart their progress toward these goals. The delegates to these Conferences of Parties (COP) have used these occasions to launch additional high-profile campaigns to reinforce the earlier commitments to conserve or restore forests.

In 2010, delegates to the tenth meeting of the Convention on Biological Diversity established twenty "Aichi Targets." These goals, to be achieved by 2020, would have halved habitat losses, thereby conserving more tropical rainforests and preserving more biological diversity. In 2011, the Bonn Challenge pledged to restore 150 mha of forests by 2020, and in 2014, the New York Declaration on Forests promised to reduce deforestation by 50 percent by 2020 and to restore 150 mha of degraded forests by 2020.[22] A fourth initiative, the Campaign for Nature launched in 2018 by the Wyss Foundation, pledged to work toward preserving 30 percent of the planet for nature by 2030. In 2021 the United Nations Environmental Program and the Food and Agricultural Organization of the United Nations declared the 2020s to be the Ecosystem Restoration Decade. Finally, at the 2021 Glasgow meeting of the IPCC, the parties approved the Leaders' Declaration on Forests and Land Use, in which governments pledged to end deforestation by 2030. To this end, governments committed US$12 billion in public funds between 2021 and 2025, and corporations committed another $7.2 billion in private sector donations. Landowners and administrators would use these funds to fight forest fires, restore degraded forests, and reinforce the rights of Indigenous peoples to forested lands.

The signatories to these pledges came largely from governments and nonprofit organizations. For example, as part of the Bonn Challenge, Mexico's National Forestry Commission, in partnership with the state government of Quintana Roo, pledged to restore 300,000 hectares of degraded forest in southern Mexico by 2020.[23] Similarly, a group of governments, companies, nonprofit organizations, and Indigenous peoples' organizations agreed to end deforestation by 2030 and announced it at the New York Declaration on Forests in 2014. The Wyss Foundation's Campaign for Nature enlisted more than a hundred nongovernmental organizations in a campaign to preserve biodiversity. With its focus on preservation and its organizational center in large, international nonprofits, the Campaign for Nature overlaps and reinforces the Bonn and New York efforts at forest landscape restoration.

Taken together, these collective actions represent a comprehensive effort across many institutions to initiate forest landscape restorations. As an enumeration of the sponsors makes clear, these initiatives came from the top down. Through declarations that set goals for changes in landscapes, representatives of influential global organizations attempted to spur global-scale restoration of forests and the associated natural climate solutions that come with the expanded forests.

Farmers, Landscapes, and Limits to the Top-Down Approaches

Progress toward these goals during the initial period after their announcement has been disappointing. The signatories to the Convention for Biodiversity failed to achieve any of the Aichi Targets in 2020 that they had set ten years earlier. The subtitle of the New York Declaration's five-year assessment says something similar: "A Story of Large Commitments and Limited Progress." Gross tree cover loss increased by 44 percent between 2013 and

2018 in the humid tropics. While high-profile actors such as the government of Norway made large-scale commitments to pay for the environmental services provided by conserved tropical forests, corporate commitments to reduce deforestation at the point of origin for their products have not diminished deforestation in ascertainable ways.[24] The companies seem unable or unwilling to exercise enough control over their commodity chains to reduce the associated deforestation. Clearly, this top-down approach to implementing forest conservation and restoration has not been sufficient to achieve the collective goals set by activists.

Initiative 20x20 represents the truism that the exception proves the rule in these narratives. Also traceable to an IPCC conference of parties, in this case in Peru in 2014, Initiative 20x20 deserves mention because, to an extraordinary degree, it seems to have succeeded at implementing projects consistent with its goals. Eighteen countries in Latin America and the Caribbean established Initiative 20x20 in 2015. It has promoted the preservation and restoration of forested lands throughout the region, using combinations of private funds, government subsidies, NGO personnel, and small landowners. While the participants in the initiatives have frequently focused on small scale projects, they have initiated an impressive range of projects across the region during the initiative's first five years. More than any of the other collective efforts, Initiative 20x20 has succeeded by linking donors, NGO personnel, public officials, and small landowners in forest conservation and restoration efforts.

Plausible explanations for the ineffectiveness of the high-profile declarations might begin with their inaccurate depictions of the ways in which landscape changes occur. As Garcia and his associates put it, "landscapes do not happen. We shape them. Forest transitions are social and behavioral before they are ecological."[25] Research on natural climate solutions has routinely ignored the human role in landscape transformations. For example, maps of potential forest cover, put together to ascertain the potential extent of reforested areas, include information on topography, soils, precipitation, elevation, and historical forest extent, but they do not include anything about human predispositions to convert agricultural lands into forests.[26] In mapping areas for potential conversion to forests, Griscom and his colleagues excluded areas with dense rural populations, ignoring the well-documented patterns of kitchen or domestic forests that small farmers create through plantings around their houses.[27] The same analysis projects the conversion of tropical pastures back into forest as the most likely, large-scale source of carbon sequestration in the near future. Reversions of pastures back into forests have occurred infrequently in the Amazon basin, and the authors of this report do not identify political or economic mechanisms that would induce this kind of land-use conversion.

In sum, models of reforestation need to begin by assuming that small farmers, loggers, and their associates are agents who can do effective things.[28] Certainly, decisions by cultivators have played central roles in recent landscape transformations. Farmers have driven processes of tropical deforestation since at least the 1960s.[29] Smallholders cleared large areas of tropical forest in Latin America during the 1960s and 1970s when programs of agrarian reform and colonization provided the impetus for land clearing. After 1980, large farmers cleared more tropical forests, and, consistent with this change, the average size of land clearings increased during the twenty-first century.[30] These associations between deforestation and the changing composition of the cultivators who clear land underscore the tight link between landholders and changes in forest cover. Farmers have made the decisions that have powered large-scale losses in forests. This destructive dynamic suggests a contrasting constructive dynamic in which changes in the incentives facing farmers fuel land abandonment and forest restoration in predominantly agricultural zones. By extension, policies that promote natural climate solutions must focus on altering the incentives faced by the people who make decisions about land use.

This "close to the people on the land" understanding of forest landscape restoration focuses on the grassroots of natural climate solutions, the landowners and users who make the decisions that restore forests on individual plots of land. Seemingly anomalous cases of forest expansion become explicable in these terms. Mexico, amid a surge in deforestation during the last two decades of the twentieth century, saw modest but still unmistakable increases in forest cover in mountainous areas of southern Mexico where community forest enterprises emerged and flourished with the assistance of the Mexican state after 1985.[31] In Colombia in 2002 three donor organizations, the World Bank, the Global Environmental Facility, and the United Kingdom's Department of Energy and Climate Change provided funds to foster intensive silvopastoral systems in the cattle ranching sector. Landholders responded positively to the incentives and supporting workshops, inducing enough regrowth of trees in the emerging silvopastures to sequester an additional 1.5 tons of carbon per year per hectare of silvopasture. The positive outcome of this initial effort led to its reauthorization in 2012 and a further increase of 90 percent in the numbers of participating cattle ranchers.[32]

The positive outcomes of these reforestation efforts, coupled with the links between local landholders and large organizations in these efforts, suggest an association between reforestation and organizational configurations that tie local to global interests. In this understanding of the ways in which natural climate solutions proceed, the agencies, featured in the New York and Bonn efforts, are one element in a set of important actors. They provide economic incentives that hasten forest landscape restoration, but other actors—Indigenous peoples, government officials, NGO activists, and farmers—also play important roles in forest conservation, restoration, and expansion. Initiative 20x20, the regional restoration program for Latin America, seems to have followed this template.

Government agencies and nongovernmental organizations do play important roles in land-use decisions, but their policies, to be effective, usually build on already existing trends in the decisions of landowners to promote forest landscape restoration in a locale.[33] Policies and donations in this context become behavioral nudges that strengthen local, place-based initiatives to restore forested landscapes. Only these types of combinations of local and extralocal interests can create political coalitions with sufficient leverage to transform landscapes in ways that realize the full potential of natural climate solutions. In this conceptualization of landscape change, coalitions of diverse actors with a common interest in forests and agriculture are the pivotal actors. They come together to create compacts, in corporatist-like processes, that obligate each member of the group to contribute in different ways to the conservation, restoration, and expansion of forests. By extension, satisfactory understandings of forest cover change must come from explanations that go up and down levels of aggregation, from small farmers who decide to abandon pastures to international donors who pledge financial support for forest restoration programs in distant locales.

The Plan for the Book

The argument in this book for a corporatist approach to understanding expansions in forest cover begins with a second, theoretical chapter that ties the large, twentieth-century declines in forest cover to a double movement, a countercoalition that arose in the late twentieth century to restore forests.[34] The success of these restoration efforts seems to have hinged on the ability of activists to construct workable, corporatist processes of decision making.

Case studies of conserved and resurgent forests in chapters 3–7 clarify the historical dynamics and associated causal mechanisms that have accompanied particular types of forest conservation, restoration, and expansion. Chapter 3 outlines the circumstances in which groups created protected areas of forest. These places avoided deforestation. Chapter 4 charts how processes of agricultural land abandonment set the stage for reforestation in Europe and the Americas. Chapter 5 describes the recent spread of planted forests and forest concessions in Africa, Asia, and the Americas. Chapter 6 examines the circumstances in which trees have repopulated cattle pastures in Latin America and grasslands in the Sahel. In widespread instances, trees have reemerged in pastures after deforestation, creating silvopastoral landscapes. Chapter 7 examines the spread of agroforests and domestic or kitchen forests in tropical locales. In most instances, small-scale cultivators create agroforests, planting fruit trees and other commercially valuable trees somewhere in the family compound or out the back door from their kitchens. Chapter 8 presents a qualitative comparative analysis of the case studies. It assesses the evidence in the case studies that corporatist political processes have provided vital supports for forest resurgence. Chapter 9 raises the analysis to the global scale, outlining how a global social movement has begun to create a corporatist-like political process to contain climate change by fostering resurgent forests around the globe.

By bringing together evidence about the circumstances surrounding the regrowth of forests and trees, this book aims to underline the empirical regularities that have accompanied natural climate solutions across diverse landscapes. These narratives of change should suggest policy directions that activists and governments could pursue to expedite and enhance natural climate solutions to global warming during the coming decades. Put differently, the analyses collected here should increase the salience of middle range, sustainability narratives, and, in collected form, contribute to a more general theory about the ways to construct sustainable landscapes.

2

• • • •

Theory

SOCIETAL TRANSFORMATIONS, CORPORATISM,
AND FOREST GAINS

About 31 percent of the earth's land area contains forests.[1] They exist in a range of different biomes: 45 percent can be found in tropical biomes; 27 percent exist in boreal regions; 16 percent can be found in temperate biomes, and 11 percent are in subtropical regions.[2] These forests vary in their densities. Humid forests have dense, closed canopies of trees. Dry forests, 42 percent of all forests by area, have open canopies with sparse stands of trees that extend over large arid and semiarid regions.[3] Forests in humid settings exhibit a dense, luxuriant growth. They sequester substantially more carbon than the sparse forests in the dry biomes.

Regional variations in rainfall have a dramatic effect on the extent of forests and the likelihood that forests, once cut, will regrow in a place. Average annual rates of precipitation for nations, grouped by region, capture the range of variation in rainfall around the globe. Central America (1,884 mm), South America (1,788 mm), and Southeast Asia (2,314 mm) are wet while North Africa and Western Asia (203 mm) and North and Central Asia (284 mm) are dry. The wetter the region, the more likely it is that seeds from nearby trees will germinate in pastures or on the floor of the surrounding forest. Given this dynamic, pastures in humid regions often have tree cover canopies that exceed 10 percent of the agricultural land area. For the 2008–2010 period, 96.1 percent of all tracts of agricultural land in Central America, a very wet region, had tree cover that exceeded 10 percent of a tract's area. In South America, another wet region,

65.6 percent of all agricultural lands had tree canopies that covered more than 10 percent of fields.[4]

Mountains cover about 25 percent of the earth's land area, and their presence influences the extent of forests, largely through the impact of mountains on agriculture.[5] Mountainous terrain complicates agriculture in ways that make it more likely that farms in mountains will fail, and in the aftermath, forests will regrow in the croplands and pastures of the failed farms. Growing seasons in mountains are short, and the transport of harvests from hard-to-access fields to markets costs farmers large sums of money. For these reasons, farmers stand a better chance of long-term economic success when they cultivate accessible, flat, and machine-friendly fields in lowland settings. By extension, the accumulating experience of farming in lowlands and highlands should gradually redistribute agriculture toward lowlands and forests toward highlands.[6]

Humans have changed these larger patterns of land use substantially since 1600. In the ensuing four centuries, humans destroyed about 40 percent of the world's forests.[7] Millions of decisions by individual farmers to clear land drove the declines in forest cover. These decisions in turn reflected local variations in access to old growth forests. Farmers with croplands adjacent to uncultivated primary forests met increased consumer demand by converting the nearby forests into croplands. Cultivators expanded the scale of their operations because it repeated what they had done before, so it was a familiar pattern of activity, and it promised to increase their household income. The timing of the agricultural expansion followed fluctuations in the prices of agricultural commodities. When prices went up, farmers converted forests into farmlands.[8]

These processes of agricultural expansion began to slow down in the twenty-first century in part because frontier forests had diminished so much in extent. Researchers could describe recently opened commodity frontiers such as the Chaco region in interior South America as one of the last remaining, largely unoccupied expanses of arable land on the planet.[9] Under these conditions, deforestation began to decline because farmers' access to old growth forests had diminished. Intact open-access forests remained, but only in more remote locations.

This global trend notwithstanding, a distinct transcontinental pattern, differentiating between the Global North and the Global South, has emerged since the 1960s. Afforestation, cropland abandonment, and forest regrowth have predominated in the Global North, while deforestation and cropland expansion have characterized the Global South.[10] Regional disparities in deforestation rates have also emerged at smaller scales. Insular Southeast Asia saw continued deforestation after 1990, while neighboring places in mainland Southeast Asia experienced forest transitions in which net deforestation gave way to net forestation.[11]

These regional disparities provide a basis for hoping that environmentally benign trends might spread to all regions, with net deforestation giving way to net forestation on a global scale.[12] In this event, a global forest transition will have occurred. Despite these hopes and instances of increased forest cover, a steady increase in deforestation rates in Africa after 2000, increased land clearing in Melanesia since 2010,[13] and an uptick in deforestation rates in Brazil after 2015 has persuaded environmentalists that the 2010s represented a "lost decade" in efforts to curb tropical deforestation and the associated greenhouse gas emissions.[14]

The Great Economic Acceleration

If forest gains are socially constructed, then it would follow that an understanding of the contemporary drivers of forestation might begin with the political and economic transformations that have driven past forest declines and, more recently, spurred countervailing forest gains.[15] A complex of political and economic changes, beginning in the nineteenth century, dramatically diminished the extent of forests.[16] Karl Polanyi captured this set of societal changes in a single phrase, "the Great Transformation."[17] It referred to interlinked, life-transforming changes in human livelihoods that industrialized economies, urbanized living conditions, increased populations, expanded agriculture, and diminished the extent of forests. After World War II, economic growth in a rapidly growing human population magnified the environmental impact of humans. John McNeill refers to this magnification of the human impact during the last half of the twentieth century as "the Great Acceleration."[18] The Great Transformation gave way to the Great Acceleration, and forest losses increased.

The Great Transformation and Acceleration also triggered an enormous increase in global economic inequality during the twentieth and twenty-first centuries.[19] By 2020, the richest nation, Qatar, had a per capita income ($117,000) 177 times greater than the per capita income ($661) of the poorest nation, the Central African Republic. Economic development spurred the increased disparities in economic outcomes. Only since the 1990s has continued industrialization and urbanization spread the wealth beyond Europe and North America. Even so, incomes and wealth remain concentrated in the more developed nations. Any thoroughgoing societal transformation in the face of climate change would presumably have to draw upon this accumulated wealth in order to accomplish its ends.

A brief narrative of the Great Transformation and Acceleration demonstrates the link between these transformations, the loss of forests, and the subsequent emergence of a social movement to save the forests. The Great Transformation began in Western Europe and spread around the globe during the nineteenth century. It transformed landscapes on all five continents. Population growth and economic prosperity in Europe spurred consumer demands for foodstuffs. European farmers responded to the increased demand by expanding agriculture. They settled and began to farm heretofore uninhabited upland settings throughout Europe during the eighteenth and the nineteenth centuries.[20] Having reached the limits of arable land in Europe by early in the nineteenth century, impoverished peasants looked for new lands to settle overseas, primarily in the Americas. State driven programs of colonial expansion facilitated the movement of poor Europeans overseas.

Waves of European settlers with associated lethal diseases eliminated Indigenous communities and established control over extensive areas in the Americas, Oceania, sub-Saharan Africa, South Asia, and to a lesser extent East Asia. The soldiers constructed colonial empires, and the merchants established trading networks between the colonies and the European metropoles. Settlers emigrated in large numbers to some of the colonies. After dispossessing Indigenous peoples from their lands, the colonists established large farms and began to ship harvests from these lands back to

European markets.[21] In newly settled regions with forests, such as the eastern United States, colonists converted forests into croplands on a large scale.[22]

The trading networks between the colonies and the metropoles persisted after decolonization, and the volume of trade expanded. European reliance on foreign lands for their foodstuffs grew during the second half of the twentieth century.[23] European merchants imported more food from Brazil and Southeast Asia. A similar dynamic emerged in China after 1980 when urbanization and industrial development led to dramatic increases in agricultural imports from Brazil and other tropical countries.[24] The resulting expansion in overseas demand for foodstuffs spurred the deforestation of old growth forests in the Americas and Southeast Asia, beginning in the 1960s.

By the turn of the twenty-first century, nations with the largest volume of agricultural exports and the fastest growing cities destroyed the most forests each year.[25] Large export-oriented farms had become the chief drivers of deforestation in these places.[26] Claimants cleared the land of old growth forests, sold the wood from these forests in urban markets, and cultivated export crops in the new fields. In many instances, investors from overseas became the owners of these enlarged landholdings.[27]

Decisions to destroy forests on a large scale began with decisions to expand agriculture. The decisionmakers brought different kinds of expertise to the task. Some worked with chain saws and started fires. They often worked as hired hands, splitting their time between the owner's large farm fields and their own nearby, smaller tracts of land. Investors and bankers provided loans to purchase livestock, machinery, and trucks. Merchants provided links to markets for the sale of products from the farms. Local officials supervised the drawing of boundaries that secured landholders' claims to tracts of forested land. Planters and landowners brought these personnel together, oversaw their use, authorized expenditures, and sometimes got their hands dirty destroying the forests. Together, the planters, workers, owners, bankers, officials, and merchants constituted growth coalitions whose activities converted tracts of primary forests into farms or plantations.

Some members of a coalition might reassemble, several years later, to convert another, nearby tract of forest into yet another large farm. All of these activities occurred in the context of a growing market for recently cleared land. In effect, the destruction of forests represented rural real estate development. The work of the growth coalitions in expanding agricultural land represented business as usual in the economic development of forested frontier regions. The participants in these coalitions engaged in a relentless pursuit of profits, so, as a group, their activities had a machine-like quality to them. They represent the engine that has driven the globalization of agriculture and the deforestation of the Global South since the 1970s.[28] Eventually, the scale of forest destruction spurred a countermovement to redress the damage done to the biosphere.

Trajectories of Landscape Change in Large and Small Farm Districts

Not all agricultural enterprises scaled up into large enterprises to meet the expanded consumer demands stemming from the Great Acceleration. This historical narrative, beginning with the Great Transformation in Western Europe, diverged in some regions of the Global South in ways that have important implications for contemporary patterns of forestation. While European settlers in some places eliminated Indigenous peoples and prepared the ground for large, settler-owned farms or land-grabbing consortia of investors, they did not or could not remove prior occupants from the land in other places, in particular in East Asia, South Asia, Mesoamerica, and parts of sub-Saharan Africa. In some instances, Indigenous peoples fled the colonizers and their associated diseases, reestablishing themselves in distant regions of refuge.[29] In these places, now known for their abundant soil resources, large rural populations, and high land prices, smallholder agriculture flourished, driven in part by skilled family labor and a related ability to outproduce large farmers per unit of cropland.[30]

In some of these regions of refuge, communities, rather than individuals, became the primary landholders. In South Asia and Mesoamerica, both regions with abundant supplies of rural labor, villagers have taken control of nearby forests, creating community forests and associated wood-based industries.[31] Villagers in these places manage their forests as a common pool resource, with the intention of supporting the villages' residents with the proceeds from the sale of forest products. These smallholder and community forests have persisted for decades despite transformations in the surrounding political economies.

These densely populated and forested areas are part of a global pattern of large farm and small farm districts.[32] Using a global compilation of agricultural census data, Lowder and her colleagues found opposing trends in farm sizes across these districts.[33] In wealthier countries, farms continued to increase in size during the late twentieth century while, in poorer countries, farms declined in size, largely through the subdivision of inherited lands. Samberg's recent research on the spatial distribution of large farms and small farms confirms the significance of the large farm—small farm divide.[34] Using MODIS-derived data on croplands and population census data for Africa, Asia, and Latin America, she and her coauthors found identifiable large farm and small farm districts. The geographical clustering of large farms in one region and small farms in other regions makes it more likely that neighboring landowners, having similar-sized farms, will make similar decisions about forest recoveries. Scaled up, these patterns suggest that identifiable regional patterns of forestation may prevail across landscapes.

Of primary importance to the arguments posed in this book, smallholders add trees to landscapes under different circumstances than do large landowners. Large landowners seem more willing to let tracts of agricultural land revert to forest, perhaps because they have more of it.[35] In contrast, smallholders and community foresters try to foster tree growth in small places, oftentimes creating "domestic" planted forests out the back doors of their homes. Disadvantaged by the small scale of their operations, smallholders and community foresters have frequently sought to diversify their streams of income from the land, especially during periods of volatility in agricultural commodity markets such as the Great Recession of 2008 and 2009. Under these circumstances, the multiple income streams from agroforestry became especially attractive to smallholders. Farmers appreciated, for example, the income from the sale of timber from the trees in the canopy of a cacao agroforest.[36] For similar reasons, small-scale cattle ranchers in the Ecuadorian Amazon allowed commercially valuable trees to sprout in their pastures, creating silvopastures. They anticipated that the sale of these trees fifteen years later would create a second stream of income for them from the land.[37] More generally, smallholders lessened their economic risks by diversifying their sources of income through the cultivation of domestic forests.[38]

A macrohistorical pattern has characterized these orientations towards risk among large and small growers. From 1960 until 2006, economic growth through globalization rewarded large-scale intensifications in land use by wealthy growers. With the sudden onset of the Great Recession in 2008, smallholders and community foresters sought to diversify their risks by investing in agroforestry and silvopastures. These activities entailed smaller shifts in landscapes, but they diversified sources of income for smallholders. Land sharing between crops and trees became more of a norm and domestic forests became more salient in household calculations. Consistent with these theoretical expectations, recent data on land use changes show a decline in the scale of changes in the tropics after 2008.[39] Given the small scale of the new stands of trees in these landscapes, only modest amounts of carbon sequestration probably occurred through these recent gains in forests. Persistent land tenure insecurity would have slowed down the emergence of forest stewardship among smallholders in these places.[40]

If decisions about reforestation differ in substance from large farm districts to small farm districts, then distinct regional forest recovery trends would presumably emerge. By extension, middle-range theories about patterns of reforestation that pertain in one region but not in another region could account for most of the observable patterns in forestation.[41] Differences in the size of landholdings and biogeographical variables such as rainfall and topography would provide the joint foundations for the differing regional dynamics that produce variable patterns of forestation from place to place.[42] The most salient of these regional dynamics that conserve, restore, or create forests are outlined in the concluding pages of this chapter.

The Double Movement: Growth in
State and NGO Power

A double movement characterized the changes in the global political economy as it accelerated during the latter half of the twentieth century. In its original theorization, the double movement took the form of a dialectic.[43] Markets expanded around the globe, and their expansion triggered a countervailing political reaction. When markets expanded, they commodified resources such as labor and in doing so stressed entire populations of workers. The low prices for commodified labor eroded the material bases for workers' livelihoods. The exploitation of workers alarmed organizers, workers, and observers. To counter the exploitation, activists mobilized people to limit the reach and influence of markets.[44] A double movement had occurred. A first dynamic, the expansion of markets, had triggered a second dynamic, a collective effort to protect humans from the ravages of the market.

The bulk of Polanyi's analysis of the double movement in *The Great Transformation* focused on labor in the industrial economy, but he made it clear that the double movement applied as well to relations between human societies and natural resources.[45] It has characterized processes of deforestation, especially in the humid tropics of the Americas. During the first stage in the latter half of the twentieth century, markets expanded spatially when governments built penetration roads into rainforest regions. Increased access to markets commodified the old growth forests in corridors along the roads. To reap profits from the now valuable trees along roads, settlers and loggers chopped the trees down and shipped them to sawmills. Corridors of destroyed forests appeared along the newly constructed roads. Landowners planted pasture grasses between the stumps in the cutover land. Within a year, cattle were grazing in these roadside pastures.

The magnitude of the forest losses alarmed ecologists and environmentalists. They mobilized in defense of the forests, and a countermovement, the second segment of the double movement, emerged after 1980. Activists in the movement campaigned throughout the world for the protection of tropical rainforests. To this end, they created numerous nonprofit organizations. They staffed environmental protection agencies and international environmental organizations. They joined innumerable local groups of birdwatchers and wildlife enthusiasts. They measured their success in terms of "avoided deforestation," forests that became parts of protected areas and Indigenous preserves. Indigenous activists, in particular, challenged the business-as-usual growth coalitions' land clearing, oftentimes in heated conversations that included threats of violence. These episodes reduced, but did not eliminate, net losses of tropical forests after 1980.

While deforestation has continued in the twenty-first century, albeit at a slower pace than previously, the capacities of states and nongovernmental organizations (NGOs) to effect change have also grown. NGOs grew in numbers and in economic capacity during the second half of the twentieth century as political activists with a common cause found each other and joined forces. A "world society" emerged, structured around norms articulated by leading activists.[46] Greta Thunberg's activism on behalf of the younger generations' interest in stemming climate change exemplifies the moral leadership coming from these groups. The Extinction Rebellion through its direct actions and the International Union for the Conservation of Nature (IUCN) through its organizational initiatives indicate the diverse ways in which NGOs have used their resources in order to achieve the collective goal of climate stabilization. These initiatives frequently bring public and non-profit institutions together in joint projects. Governments and NGOs, jointly and separately, initiated 467 REDD+ (Reduced Emissions from Deforestation and Degradation) projects around the world between 2007 and 2018.[47] The sheer number of REDD+ partnerships during this period indicates how vibrant the double movement NGO–government sector has become.

These increased state capacities often occurred in the aftermath of climate change induced, extreme weather events. These shocks provoked regime shifts in which new ecological conditions pushed a community past a critical threshold, changed economic conditions, and altered political incentives, creating crises that in some instances spurred government attempts to reforest landscapes.[48] For example, during the Depression of the 1930s, the shocking images of drought in the Dust Bowl refocused political agendas and increased popular support for initiatives such as the Soil Conservation Service that in some locales led to forest recoveries.

The growth in these state capacities, although highly uneven geograph- ically, made it more likely that states would initiate efforts to restore degraded landscapes and in doing so promote natural climate solutions. State activism in defense of nature grew demonstrably after 1992 when leaders from the Global North and South met in Rio de Janeiro for the Earth Summit. This meeting launched two United Nations–sponsored efforts to curb environmental destruction, the Convention on Biological Diversity and the Framework Convention on Climate Change. These UN initia- tives represent two high profile organizational efforts that appear to have increased the efficacy of recent initiatives to conserve, restore, and expand forests.

Societal Corporatism: The Politics of Forest Conservation, Restoration, and Expansion

In the larger context set by the market driven Great Acceleration, the dynamics that historically generated new forests often seemed to arise out of direct interactions between trends in markets and the environment. For example, downward trends in the prices of crops after World War I bank- rupted farmers on marginal agricultural lands in the United States. These farmers abandoned their agricultural lands shortly thereafter, and old field succession began, leading in thirty years to new secondary forests. While this economic account of the origins of secondary forests may explain the increase in some secondary forests during the nineteenth and twentieth centuries, it misses a crucial political dimension to many of the events that have generated new forests since the 1980s.

This empirical shortcoming has its origins in the liberal understandings of market-based societies that became popular with the emergence of the Great Transformation and Acceleration during the nineteenth and twenti- eth centuries. Liberal conceptions of societies feature individuals who trade in markets and states that guarantee the rights of individuals. Émile Dur- kheim pointed out an empirical problem with these conceptions of societ- ies. They neglect the intermediate groups that stand between and medi- ate relationships between individuals, markets, and the state. Guilds and churches represented the most influential intermediate groups during the medieval era. Nongovernmental organizations created by businesses, laborers, churches, and racial or ethnic groups have become the most common intermediate groups during the modern era. These corporate groups articulate codes of conduct that represent the sinews of society, both locally and globally. In Durkheim's words, "a nation can be maintained only if, between the state and the individual, there is intercalated a whole series of secondary groups near enough to the individuals to attract them strongly into their sphere of action, and, in this way, drag them into the general torrent of social life."[49]

These intermediary groups appear to play pivotal roles in recent narratives of forest cover gains. They convene organized gatherings of representatives from different groups of people with a common interest in forests. The assembled people have ties to one another through their common interest in landscapes. In some instances, these people can be considered a unified body of individuals, a corporate grouping. The groups draw on diverse peoples: Indigenous groups, farmers, environmentalists, donors, and government officials. Through processes of negotiation, the representatives of these peoples create internal compacts that increase the likelihood of forest cover gains. In some instances, the diversity of represented groups in these compacts lays the groundwork for challenges to the domination of local politics by forest-destroying growth coalitions. In narratives of resurgent forests, these corporatist political entities represent proximate, as opposed to underlying, causes for forest cover conservation, restoration, and expansion.

The presence of intermediary groups who participate in corporatist processes within the landscapes sector has been contingent historically. Before 1980, groups in the forest—interested organizational field had little contact with one another. Forest-dwelling Indigenous groups did not have NGOs that represented them in government and in governance fora. Potential donors of funds for the protection of forests had few, if any, ties with Indigenous groups. Environmentalists had relatively few local or international NGOs. Scientists had yet to take advantage of the new remote sensing technologies, and they had yet to form interdisciplinary interest groups to study processes of land use change. Many states did not have environmental ministries. The leaders of private enterprises made commitments to eliminate deforestation from their production processes without much knowledge or firsthand experience with the suppliers of raw materials for their finished products.[50]

Departments of agriculture did exist, but most farmers in the Global South, especially in locales where smallholders predominated, did not have titles to the land that they cultivated. They may have had a bill of sale for their land, but they had not received a state-issued title that recognized their ownership of it. Under these circumstances, they were vulnerable to land invasions. Without titles, they faced bleak prospects when they sought redress in the courts after a land invasion. Female smallholders faced especially difficult prospects.

In this kind of fragmented system, the coalition building necessary to implement environmental protection became almost impossible. Without these political supports, efforts to conserve or restore forests frequently fell short. This record of ineffectual efforts began to change after the 1980s as intermediary groups have proliferated across the landscapes sector. With this change in the forest-interested organizational field, the prospects for collective action to conserve, restore, and expand forests have improved. When collective action has occurred, it has most frequently taken a corporatist form.

The agents of corporatism organize political life around occupations or sectors of an economy.[51] The sectoral associations bring together public officials, businessmen, and labor leaders from a particular sector such as manufacturing. Representatives of the rich and the poor encounter one another in these negotiations. The representatives of these groups negotiate sector specific agreements about wages, taxes, and government assistance that apply to participants in the sector. The negotiations that lead to the compacts create links between different groups in a sector of society, thereby counteracting fragmentation in the sector and increasing the efficacy of political initiatives by actors from the sector.

The compacts stress "concertation" in which policies in one part of a sector, businesses for example, integrate with policies in another part of a sector, labor for example. To make these agreements acceptable to all parties, negotiators try to arrive at agreements that spread gains and losses across a sector. Political exchanges occur between the parties negotiating the compacts.[52] Austria and the Scandinavian countries exemplified corporatist political economies in the post–World War II era in Europe. Schmitter refers to the post–World War II democratic regimes in Europe as "societal corporatist" systems.[53] The unified, sector specific decision-making processes of the post–World War II European societies had a democratic dimension because they brought together base organizations and elite groups in negotiations that culminated in the compacts described earlier.

Corporatist polities differ from pluralist polities in the way that they represent economic interests. Pluralist regimes promote political organization on the basis of common material interests, so citizens in pluralist regimes will typically organize themselves into a business owner party and a working-class party.[54] In a corporatist polity, the sectoral associations bring employers and laborers in a sector together to negotiate compacts in which the different interest groups concert their actions in the sector. In this respect, corporatist polities muffle or mute class-based conflicts.

Studies of societal responses to political—economic turmoil created by business cycles have concluded that corporatist polities weathered these political-economic storms better than pluralist political economies because the consensual corporatist processes enabled collective action that addressed the deprivations that came with the economic turmoil.[55] These dynamics suggest a more general affinity between societal crises and corporatism. It is probably not an accident that corporatist policymaking processes increased in number when states mobilized to fight each other at the outset of World War I.[56] If, as expected, the turmoil associated with climate change accelerates in the next several decades, the ability of corporatist political processes to foster robust, climate stabilizing programs of collective action should increase the appeal of corporatist political processes to policymakers.

Corporatist processes occur in state as well as societal forms.[57] In several notorious examples in Europe during the 1930s (Germany, Italy, and Spain) authoritarian regimes instituted corporatist processes in efforts to extract a greater collective effort from their citizens. Contemporary China resembles this political form in crucial respects. In Philippe Schmitter's terms, China represents a "state corporatist" system. State corporatist regimes would have more centralized decision making than societal corporatist regimes.

The emphasis in corporatist systems on integrating disparate groups into policymaking processes made it more likely that the interests and welfare of all major groups would be served by the compacts. More generally, corporatist strategies have strived to overcome fragmented political systems. In so doing, they enable collective action. Societal corporatist regimes would include a wide range of participants in decision making, embracing a kind of procedural equity. The compacts would also, in substance, endorse a distributional equity in which the different parties only agree to roughly equitable outcomes. The compacts, in substance, would endorse changes in norms, in how things are done. The changes would range from norms in governments about how to raise money to norms among workers about how to make tools or cultivate a piece of land.

Applied to changing landscape ecologies, societal corporatist regimes would organize the forest–agriculture sector with a wide range of participants drawn from farming, forestry, and grassroots NGOs as well as from overseas donors, government officials, and activists from environmental and Indigenous NGOs. These decision-making organizations bring concentrations of wealth from donors in the developed countries and power from government officials into contact with grassroot organizations of large and small landowners who make decisions about land uses. Through negotiations, sector participants would hammer out compacts which oblige each of them to a particular course of action. The donors might pledge monetary support, but they would only fulfill their pledges if the state addresses and resolves, with smallholder participation, the tenure insecurities that plague landholders in so many rural areas. Forest owners would then pledge to leave their trees standing. The donors, the officials, and the farmers would implement a cash-on-delivery system with payments for trees.[58] All of these changes could only occur in places where the negotiators share a belief in the need to respond to the climate crisis through changes in the norms that govern how people use land.

The hallmarks of corporatist-controlled processes of landscape transformations would be as follows.

- Multiple parties from different strata in society, from international bankers to local politicians and farmers, confer about increments in forest cover in large and small locales.
- The conferring parties agree to compacts in which some transfer funds in return for the other's conservation and restoration of forests. The parties concert their activities.
- These parties meet repeatedly over the years, with some changes in the composition of the negotiators, to construct and implement successive plans for forest conservation, restoration, and expansion.

Over time, the repeated meetings, the compacts, and the concerted activities create a corporatist polity. The resulting integration of the disparate groups in the forest—agriculture sector would increase the likelihood that commitments such as the Bonn Challenge and the New York Declaration would produce substantial increments in forest cover and carbon sequestration. Global accords would cascade down the societal scale to local agreements.[59] The inclusion of overseas donors in the negotiations increases the probability that the compacts would bridge and maybe even attenuate the enormous economic inequalities between the Global North and South at the same time that they would produce substantial increases in carbon sequestering landscapes.

Participants in a landscapes sector would share a common interest in forests and agriculture, but they come to their shared interests from quite different backgrounds. Farmers and loggers have had long-term foci on earning livelihoods from the land, and they approach forested land as a natural resource that can earn them a livelihood. Representatives from international environmental NGOs have paramount concerns with the elevated global rates of greenhouse gas emissions and the biodiversity impacts of deforestation. The inclusion of representatives with global interests in these negotiations compels the creation of compacts that internalize the externalities that result from forest destructions and, in so doing, advance the provision of a global public good such as a stabilized climate and biodiverse flora.[60] Compacts in sectors with this combination of interests would in theory prevent the wholesale, climate changing destruction of forests at the same time that they remunerate farmers for their efforts to restore forests.

With a wide array of participants, the negotiators in a sectoral organization could go up and down geographical scales or levels of aggregation in their discussions and in the policies and practices that they choose to advocate and implement. Out of a succession of meetings, a political program emerges in which resources from the concentrations of wealth in affluent centers finance forest restoration in impoverished peripheries. The compacts reflect the characteristics of the preceding discussions. The forest-enhancing projects, like the preceding discussions within the landscapes sector, reach across scales from local communities to the global arena.

For these corporatist entities to succeed in reducing deforestation and encouraging reforestation, they must incorporate important actors such as farmers from the forest-destroying growth coalitions. Vocational groups such as farmers wield substantial political power in making compacts about forests because they shape the norms that govern land uses, in particular norms about deforestation but also norms about land abandonment and norms about tree planting. They have representatives in the forest—agriculture sector negotiations who have mandates to promote the interests of farmers on issues such as land tenure.

These joint efforts of donors, officials, and landowners presume a modicum of unity among the participants in a sector.[61] They have become participants in the double movement committed to restoring or conserving forests. Participants in the negotiations around the compacts would share, for example, a commitment to climate stabilization and sustainable agricultural development. Elements in the compacts would serve various ends. For example, land tenure reforms, by securing smallholders' titles to land, would encourage long-term perspectives about farms which in turn would promote sustainable agricultural development and enable carbon offset payments to farmers.

The inclusion of diverse peoples within the organizational frame of corporatism explains why corporatist structures associate so consistently with policies that succeed in expanding forest cover. In this respect, societal corporatism provides an antidote to the top-down approaches to forest restoration exhibited in the Bonn Challenge and the New York Declaration on Forests. In corporatist compacts, altruistic environmentalists, so much in evidence among the elite signatories to the top-down declarations, make common cause with the defensive environmentalists, embodied by Indigenous peoples.[62] Corporatist approaches connect distant elites committed to change in landscapes, through facilitating intermediaries, to local landholders in positions to make changes in landscapes. Initiative 20x20 takes this form. There are good reasons to believe, based on the first few years of these transnational alliances between, for example, Norway and local groups in Tanzania, that these alliances will become recurrent ones, with an experience with the first project laying the groundwork for additional projects at a later date.[63]

Two organizational initiatives introduced in the twenty-first century, the United Nations–endorsed REDD+ initiative and the array of ecocertification schemes in agricultural communities, provide recent examples of corporatist-like processes of decision making under construction. While both of these efforts remain works in progress, they promise significant increments in forested landscapes. The following paragraphs provide summary accounts of these corporatist efforts to enhance forests.

Reductions in Deforestation and Degradation: REDD+ Projects

Since the early 2000s, NGO staff, community organizers, and scientists have tried to overcome fragmentation in the forest sector by constructing multitiered systems of payments for environmental services (PES), identified by the acronym REDD+.[64] Beginning with small, experimental programs in two countries in the late 1990s, REDD+ became "the leading option for climate change mitigation in the forest sector."[65] The REDD+ network links donors and purchasers of carbon offsets in wealthy countries, through governments in forest-rich countries in the Global South, to landowners whose forests extract carbon from the atmosphere and store it in trees above and below ground. In these arrangements, in theory, companies in the wealthy nations purchase carbon offsets from the governments of landowners who sequester carbon. The companies purchase the offsets in order to achieve a stated goal of carbon neutral operations or to comply with government regulations that mandate reductions in carbon emissions. Governments in the forest-rich recipient countries use the funds from the companies to pay landowners and communities for sequestering carbon in their trees. In this manner, participating landowners and communities have an incentive to allow trees to grow on their lands. By linking the off-set purchasing companies through governments in the Global South to landowners in the Global South, REDD+ like arrangements knit together, at least theoretically, the participants in the landscapes sector. Put differently, REDD+ schemes link organizations in wealthy carbon-emitting societies to landowners in the poorer, sometimes carbon-absorbing societies in the Global South.

While the organizational architecture of REDD+ would seem to create an economic mechanism for sequestering large amounts of carbon in forests, it has in practice not done so in most instances. Two problems that reflect fragmentation in the landscapes sector have prevented the implementation of REDD+ programs. First, the demand for carbon offsets by affluent consumers of fossil fuels has been weak, largely because the wealthy countries have done little, to date, to mandate reductions in greenhouse gas emissions. Under these circumstances, companies that have had difficulty reducing their emissions have not had to purchase carbon offsets to avoid fines for failing to curb their emissions. As a result, the prices of the offsets have remained low, and the administrators of REDD+ have had only small amounts of money to offer as payments to landowners for environmental services such as carbon sequestration.[66] Second, insecure land tenure in rural districts of the Global South has prevented the implementation of many REDD+ agreements. To receive payments for carbon sequestration in forests, landholders need proof that they own the land where the carbon has been sequestered. In countries with the most carbon sequestering potential in the Global South, Brazil, Indonesia, and the Democratic Republic of the Congo, most landowners do not have secure titles to land, so they do not qualify to receive payments for carbon sequestration through REDD+ programs.[67]

Titling initiatives in several countries give some grounds for hope about the eventual impact of REDD+ on rates of carbon sequestration. Massive titling efforts in Peru, Brazil, Indonesia, and other countries may gradually dismantle the insecure tenure impediment to accomplishing REDD+. Activists in Tanzania have had a largely positive experience with REDD+, in part because Tanzania, going back into the twentieth century, invested heavily in tenure clarification in rural areas, so they began the REDD+ experience having already achieved secure tenure. The experience in Cameroon suggests an alternative path to implementing REDD+. In this instance, REDD+ projects began in a context of generalized tenure insecurity, but efforts by the REDD+ implementing organizations in villages improved smallholders' tenure security which in turn made smallholders more willing to participate in REDD+ programs.[68]

Where successful, REDD+ projects initiated a political process that linked affluent actors in the Global North, through intermediaries, to impoverished actors in the Global South. Compacts between these parties promise to sequester substantial amounts of carbon in tropical landscapes in return for payments by wealthy entities for carbon offsets. In their general political configuration, implemented REDD+ projects worked through corporatist political mechanisms.

Ecocertification

REDD+ agreements promote natural climate solutions by exerting direct control over land uses in defined territories. Ecocertification schemes promote natural climate solutions by indirect means. Certifiers distinguish between flows of commodities grown under different conditions, some more ecologically friendly than others.[69] Those commodities grown in more forest friendly conditions would qualify for sale in markets patronized by consumers willing to pay higher prices for commodities certified to have been grown in sustainable landscapes. Intermediary organizations such as the Rainforest Alliance or the Forest Stewardship Council certify that farmers grew the commodities for sale in the high-priced markets in a sustainable manner. Ecocertification schemes first appeared during the 1990s. Certified organic coffee is probably the best known of the ecocertified foods. Between 2005 and 2015, ecocertified growers multiplied in the coffee growing regions of the tropical world.

Ecocertification schemes exhibit several important features of corporatist governance. Through intermediary organizations, they bring together wealthy consumers, certifiers, and producers, large numbers of whom grow their crops on small landholdings. These parties negotiate compacts over the use of land for limited periods of time. These compacts between wealthy consumers and small-scale farmers made the certification business a possible way to redistribute wealth while pursuing an environmental agenda.

Like other corporatist arrangements, ecocertification also involves repeated negotiations from year to year about the substance of a compact, in this instance the land use practices required in order to qualify a product for sale in a preferred market. The history of land use on certified lands has become a contentious issue with some commodities. A history of recent deforestation did not disqualify oil palm planters from certification as sustainable oil palm plantations during the 2005–2010 period. A decade later, the early deforestation records of these planters had become more controversial and the subject of negotiations about certification between producers, certifiers, and consumers.[70]

In these respects, the negotiations between rich and poor, the economic exchanges between negotiating parties, and the repeated nature of the negotiations all represent essential elements of corporatist schemes of governance. They apply to ecocertification schemes as much as they do to REDD+ projects. The prevalence of these corporatist arrangements in forest saving efforts suggests that corporatist political processes have become an integral component in forest-saving efforts.

Corporatist Paths to Forest Conservation, Recovery, and Creation

While corporatist political processes have characterized most recent instances of forest resurgence, the human ecological form of forestation has taken two distinct paths. The first path, historically, saved forests by enclosing them in protected areas. In these instances, a protected forest continues to grow and accumulate carbon above and below ground. A protected

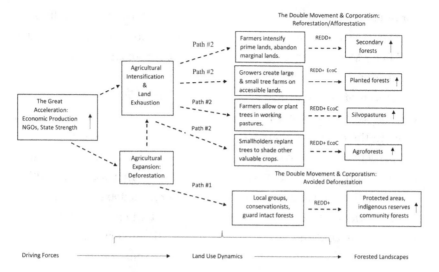

FIGURE 2.1 A causal model. Drawn by the author.

forest may not have increased in extent, but it avoided deforestation. A second path, more common in recent years, restores forests in degraded and deforested landscapes. The two paths do not represent real alternatives to one another. The first path, focused on protected areas, seems most feasible in places with intact forests. The second path, focused on restoration, seems feasible in disturbed or degraded landscapes where additional human interventions might actually increase the environmental services coming from the land. Figure 2.1 and the following pages outline the two paths to forest gains.

Path 1: Intact Forests, the Conservation Movement, and Avoided Deforestation

The acceleration in rates of deforestation during the 1960s and 1970s had by the late 1980s mobilized conservationists into countercoalitions whose leaders exerted political pressure for the conservation of intact rainforests.[71] Like other coalitions in corporatist polities, this one had a diverse membership. Often these coalitions included Indigenous peoples who advocated for reserves where they could continue to pursue forest-based livelihoods. The countermovement also included wealthy donors such as the Gordon and Betty Moore Foundation, which bought tracts of land and established them as protected areas.

A series of conflicts over deforestation-inducing development projects and the creation of protected areas to deter deforestation contributed, over time, to the creation of durable conservation coalitions. The diversity of the participants in the coalitions, donors, Indigenous peoples, government officials, and activists from environmental NGOs, increased the political appeal of the compacts that established the protected areas. The long list of signatories to the declarations at Bonn in 2011 and in New York in 2014 indicates the continuing political appeal of the protected area countermovement. The diverse sources of support for some of these protected areas indicates how countermovements to save the forests can evolve over time into a corporatist political process that brings diverse groups together repeatedly to negotiate compacts that meet the participants' needs and preserves forests.[72]

Over time these mobilizations have succeeded in placing restrictions on the cutting of a growing proportion of the world's intact forests. Between 2000 and 2020, the proportion of the world's forests in protected areas increased from 14.12 percent to 17.81 percent.[73] By 2020, 28 percent of Brazil's forests grew in protected areas, and another 31 percent of the country's forests grew inside Indigenous reserves.[74] Public officials, such as those in Brazil, added to these efforts, at least in some locales, by enforcing laws that prohibited the clearing of more than 50 percent of the land on farms.[75] When these combined efforts succeeded in expanding protected areas, the countermovement avoided deforestation.

Forest preserves differ in the degree of protection that they enjoy. Some forests are only largely intact. Community forests in South Asia and Mesoamerica have avoided deforestation, mostly through the efforts of nearby communities whose members manage the forests for their livelihoods. While local residents harvest products from the forests, they do so as stewards who limit human interventions in the forests in the interests of preserving them and the villagers' livelihoods for the near future. In some instances, these forests have avoided deforestation at the cost of experiencing some frequent, small-scale harvesting.

Newly protected forests only continued to preserve biodiversity and sequester more carbon when the newly designated preserves prevented business-as-usual coalitions from clearing forested land within the boundaries of protected areas. Government guards for protected areas, Indigenous patrols of preserves, and surveillance by community foresters all oppose and disrupt land clearing by members of the growth coalitions. In doing so they reduce the damage done to old growth forests by the business-as-usual coalitions in rural areas.

A recent, global-scale analysis of the effectiveness of protected areas indicates through a matching analysis that protected areas reduced deforestation rates by 41 percent compared with similarly situated unprotected areas. The staffs of large protected areas proved less effective at preventing deforestation, in part because they had, per unit area, fewer resources available to protect forests. Wealthier nations provided more protection through their parks because, presumably, they spent more money on the protection of each park. Protected areas promote the provision of natural climate solutions. Given these effects, a case could be made that financial support from international organizations should be made available to maintain protected areas in poor countries.[76]

Path 2: Degraded Lands, Intensified Agriculture, and Resurgent Forests

While one path to resurgent forests preserves intact forests, another increasingly common path rehabilitates degraded forests, restores destroyed forests, and establishes new forests. This restoration path begins in a predominantly agricultural landscape. By early in the twenty-first century, humans had destroyed approximately one half of the forests in the tropical biome.[77] These agricultural landscapes contained mosaics of degraded forests, scrub growth, grasslands, and still productive croplands. Farmers working these lands routinely resorted to new agricultural techniques to maintain soil fertility and bolster the size of their harvests in a context where urban consumers made escalating demands for agricultural commodities.

The earlier expansion of agriculture in long deforested places had also created over time a growing bank of exhausted lands that could not be restored to productive capacity without extensive investments in new technologies. This historical sequence of deforestation followed by land degradation implies that agricultural intensification efforts should have grown in prevalence over time as the fertility of some cleared lands declined to the point where they could no longer be farmed. Intensification of these lands often required expensive new inputs and credit from third parties. New technologies become an important means to higher yields under these circumstances. This dynamic was especially evident in long-settled regions with extensive areas of degraded land. Farmers raised yields by applying new inputs, herbicides or mineral fertilizers for example, to the land.[78] This general situation resembled the circumstances outlined in Ester Boserup's *Conditions for Agricultural Growth* in which population growth, with its increases in demand for foodstuffs, spurs the adoption of new techniques in agriculture.[79] Here the increases in demand came from increases in the affluence of consumers as well as from increases in population. The agricultural intensification in response to soil exhaustion and increased consumer demand led to increments in forest cover in at least four different ways (see path #2 in figure 2.1).

First, amid the tremendous agricultural expansion of the nineteenth and twentieth centuries, farmers converted terrain of all sorts into fields for cultivation. Many of the newly cultivated fields proved to be ill suited for farming. Periodic declines in the prices of agricultural commodities made it impossible to profit from the meager harvests from these lands, so farmers abandoned these lands in increments, tract by tract. Additional production from intensified agriculture on the more fertile fields kept the prices of staple crops such as maize, wheat, and rice low enough to bankrupt disadvantaged farmers. Under these circumstances, poor farmers in disadvantaged locations abandoned agriculture altogether.[80] More prosperous farmers experienced the same dynamic, except that they only abandoned their least productive fields. These well-off farmers often, at the same time, intensified production on their more fertile fields, thereby driving the prices of agricultural commodities even lower.[81] Climatic conditions permitting, the abandoned croplands then reverted spontaneously over the course of several decades to secondary forests. Governments, most notably in the United States during the New Deal, created financial incentives to hasten land abandonment. Franklin Roosevelt's "brain trust," working in tandem with the newly created Soil Conservation Service and participating farmers, created a corporatist-like regime to hasten the reversion of eroded croplands into forests.[82]

Second, landowners began to plant trees on degraded lands. Rising prices for timber, driven by increases in consumer demand for wood during an earlier era, had persuaded timber companies to log more remote primary forests. The extension of logging operations into distant forests had raised the costs of transporting the logs to markets to the point where landowners and merchants began to consider planting trees in close proximity to cities, with the intention of saving money on transporting the harvested wood the comparatively short distances to markets. Landowners responded to these price signals by creating forest plantations in accessible locations.[83]

Many of the newly planted forests replaced scrub growth left by previous land users. Given the preexisting scrub growth, forest plantations represented intensified land use at these sites. In this sense, the concentration of plantations in accessible sites along coasts or in places with dense networks of roads represented a new land use erected on the ruins of previous land uses. States, in partnership with large and small plantation owners, created infrastructure such as roads that accelerated the creation and expansion of the plantations. The concerted increases in market access, reforms in land tenure, and initiatives from farmers resembled a corporatist process, especially in authoritarian polities such as Vietnam.

A third dynamic has taken hold in grazing areas of the Global South where ranchers and pastoralists have established and maintained pastures. As the pastures have aged, their productivity has declined, and farmers have sought additional sources of income from the land. By allowing commercially valuable trees to sprout in pastures, pastoralists and small-scale cattle ranchers have promoted the emergence of silvopastoral landscapes with some capacity for carbon sequestration. The sale of wood from these trees in pastures created a second stream of income from the land for small-scale ranchers. In effect, the trees in pastures represented an intensified form of land use. In some settings, such as Niger in the Sahel, external assistance and tenure reform spurred farmers, in a corporatist process, to plant trees around their homesteads.

Fourth, the appeal of trying to restore degraded lands grows in old colonization zones as alternative agricultural economic opportunities diminish over time. Areas on Sulawesi have appeared where the remaining forested areas on the island are protected and the formerly forested areas have now become degraded grasslands dominated by invasive grasses such as *Imperata cylindrica*. Under these circumstances, a restorative dynamic driven by technological changes can take hold in a locale. In the example of Sulawesi, farmers sought to replace the *Imperata* grasslands with a cacao predominant agroforest. This restorative effort became possible only through an extensive and innovative use of herbicides to eradicate the *Imperata*.[84] In doing so, farmers prepared the ground for the replanting of cacao and the associated shade trees. Agricultural intensification occurred, and it increased the extent of agroforests in a rural district. These reforms put the smallholders in a position to benefit from ecocertification schemes that link environmental quality-conscious consumers with smallholder producers. In some of these circumstances, a corporatist organizational field, populated by companies, NGO representatives, and small farmers, emerged around ecocertification.

A cross-tabulation of reliance on natural regeneration or plantings on the one hand and trends towards more large landowners or more smallholders in districts helps to situate these paths to forest recoveries on already cleared land (see figure 2.2): (1) Regrowth in old fields following land abandonment, (2) planting trees in forest plantations, (3) spreading silvopastures, and (4) increases in domestic, kitchen-oriented agroforests. Table 2.1 provides estimated rates of carbon sequestration for these land use transitions, primarily in the tropics. For purposes of comparison, when the land use transition consists of invasives, like *Imperata*, that spread over already degraded croplands, no carbon sequestration occurs ($-.2$ t C ha^{-1} yr^{-1}). Young (less than twenty years old), naturally regenerated secondary forests showed the highest rates of carbon sequestration. Shade grown coffee groves

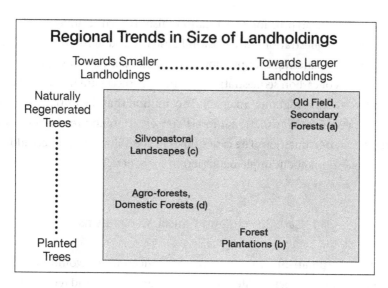

FIGURE 2.2 Cross-tabulation of reforestation: Size of landholding by reforestation pattern.

TABLE 2.1 Carbon Sequestration Rates—Varieties of Reforestation

Land use transitions from permanent cropland to:	Carbon uptake (t C ha⁻¹ yr⁻¹)	Source
Imperata—tropics	−0.2	6th Assessment, IPCC*
Natural regeneration—temp. and tropics	9.1–18.8	Bernal et al. 2018
Shifting cultivation, short fallow—tropics	3	6th Assessment, IPCC*
Shifting cultivation, long fallow—tropics	7	6th Assessment, IPCC*
Silvopasture—tropics	1	McGroddy et al. 2015
Sahel, parklands—tropics	1.1	Nair et al. 2009
Tree farms, exotics—tropics	5	Lewis et al. 2019
Coffee agrofor—tropics	6.3	Nair et al. 2009
Cocoa agrofor—tropics	1.5	Lewis et al. 2019
Rubber agrofor—tropics	3.7	Lewis et al. 2019
Oil palm—tropics	6.1	Pulhin et al. 2014
Domestic, "kitchen" forests—tropics	4.3–8.0	Nair et al. 2009

*Land Use, Land-Use Change and Forestry, IPCC, https://archive.ipcc.ch/ipccreports/sres/land_use/index.php?idp=184

and acacia–eucalyptus forest plantations also showed relatively high rates of carbon sequestration. Silvopastoral systems and agroforests with few trees in the sheltering canopy showed lower rates of carbon sequestration. These rates of carbon sequestration presume the successful implementation of reforestation programs, an enabling condition that the initial, pessimistic reports of the transnational forest conservation and restoration campaigns call into question. The case studies in the following chapters address these questions about implementation in more detail.

Case Studies and Causal Mechanisms

Societal corporatism, with its focus on concertation, would encourage simultaneous changes in international assistance, state land tenure bureaucracies, and landholder forest conservation practices. These changes, together, would, in theory, enable a full range of natural climate solutions, including new protected areas, regenerated secondary forests, silvopastures, seedling subsidies for small scale forest plantations, and rejuvenated agroforests. If this theory finds empirical support, similar types of coalitions would characterize successful programs to establish protected areas, create Indigenous preserves, promote land abandonment, establish forest plantations, spread silvopastures, and support a further expansion of agroforests. An absence of corporatist processes would characterize efforts that failed to produce widespread expansions in forest cover or biomass. Under these circumstances, the success of localized efforts at forest conservation and restoration would hinge on broader institutional changes, in particular a resort to corporatist processes of decision making.

These theoretical suppositions, to be credible, require some empirical confirmation. Empirical support for this argument about the importance of corporatist political processes would come from case studies of gains in forest cover and biomass. Through detailed narratives of events, sometimes referred to as process tracing, case studies outline the causal mechanisms that have contributed to increases in forests.[85] The inclusive criteria for data in case studies also makes them appropriate for instances of conjunctural causation, likely here, where political, economic, and biophysical variables all play a role in increasing forest cover.

Two sets of criteria shaped the choice of case studies. First, the archive for each case study had to contain a fine level of detail, down to data on individual farmers. Otherwise it would have been difficult, if not impossible, to ascertain whether or not a corporatist process had triggered the expansion of the forests. Second, the case studies in the aggregate had to show variable amounts and types of forest gains as well as variable political processes. An analysis across the assembled case studies would then make it possible to assess the degree of association between political processes and forest cover change. The absence of an association between corporatist political processes and increased forests would disconfirm the hypothesis about the outsized influence played by corporatist processes in achieving forest gains.

The initial assessment of evidence comes from case studies of different types of forest resurgence. The following five chapters focus on different types of forest gains that represent, in one way or another, natural climate solutions. These chapters have a common organizational frame. They begin with estimates about the global extent of this type of forest gain. The case studies of this type of forest gain follows. They describe particular interactions between political, economic, and ecological conditions that did or did not produce the gains in forests and forest cover. Summaries at the end of each chapter address questions about the magnitude of the gains in forests that would occur if the process under study became widespread.

3

• • • •

Forest Losses, the Conservation Movement, and Protected Areas

For millennia, forests in the Global South enjoyed a kind of passive protection. The relatively small numbers of humans and the associated small scale of human enterprises protected most forests from wholesale destruction. The accelerated growth in human numbers and enterprises in the twentieth century ended the protection conferred on forests by distance and slow means of transport. Roads became the visible and tangible expression of the Great Acceleration in the human enterprise following World War II. Roads and the associated vehicles broke down the barriers of time and distance that had protected forests from expanding human enterprises. Aspiring farmers came with the roads, and they quickly cleared roadside lands, creating corridors of destroyed forests along the roads. Fields for crops and pastures for cattle replaced the forests.

The rapid loss of old growth tropical forests, a first movement, spurred a second, reactive movement to protect forests. As alarm over species extinctions mounted with the loss of forests, conservationists expanded their goals for the extent of protected areas. By 2015, E. O. Wilson and others had begun to argue that protected areas should encompass one half of the planet.[1]

By 2020, the world had 240,000 protected areas, covering about 15 percent of the world's surface. The Global South had more extensive protected areas than the Global North, with 31 percent of South America's forests and 27 percent of Africa's forests protected. Brazil had the world's largest network of protected areas. The large extent and rapid increase in preserved areas testifies to the effectiveness of the second, countervailing movement. The greatest gains in protected areas between 1992 and 2015 occurred in tropical rainforests whose destruction had touched off a wide-ranging effort to protect the biosphere.[2] The creation of protected areas in Ecuador where the Andes meet the Amazon illustrates this dynamic.

A Case Study: Deforestation and Protected Area Expansion in Ecuador's Eastern Andes

For centuries, the dramatic landscape where the Andes meet the Amazon in Ecuador contained luxuriant rainforests and a small scattering of people, the Shuar, an Indigenous group, and a few Spanish-speaking, mestizo migrants from the Andean highlands. The state claimed all of the land, but in fact the land was open access, there for the taking. Few people had legally defensible claims to land. Land use changes accelerated in this region during the 1960s when the Interamerican Development Bank decided to finance the construction of a road, the Carretera Marginal de la Selva (highway at the edge of the forest), at the eastern base of the Andes. The plans for road building stirred the interest of would-be migrants to the region. Mostly landless Spanish-speaking peasants from the Andes, the migrants followed mule trails and newly constructed stretches of the Carretera Marginal down into the rainforest region. Once there, they claimed forested tracts of land along the planned route of the road. To make their presence on the land visible to all passersby, the peasants cleared portions of these forests for cattle pastures.

The rapid deforestation along the projected corridor of the new road alarmed an advisory committee of natural scientists from the World Wildlife Fund. They feared that the land clearing would threaten the rich biodiversity of the forested Andean foothills. Photographs such as figure 3.1 corroborated these fears, with its heaps of dead underbrush and felled trees in a destroyed tract of old growth rainforest at the base of the Andes. To prevent further settler incursions into this biodiversity hotspot, the Ecuadorian state followed the recommendations of the alarmed scientists and

FIGURE 3.1 Photo of destroyed vegetation, 1969. tropical forest, Ecuadorian Amazon. Photo by Gary Hill.

created the Sangay National Park in 1976. The new park extended from the Amazon plain up into the Andean highlands and included active volcanos at the highest points of the Cordillera Oriental.

Coincidentally, the exploitation of new oil fields in northeastern Ecuador had filled the coffers of the central government with tax revenues from oil exports. These revenues eased the financial burdens of establishing and maintaining the new park. Two decades later, after Ecuador had received unwanted attention as the Latin American country with the then highest rate of deforestation, the central government doubled the size of the Sangay National Park by extending its boundaries southward into a rugged, virtually uninhabited region of the Cordillera Oriental. The absence of people in the newly included southern reaches of the park facilitated its expansion by reducing local opposition to the enlarged park.

The new park and the recent settlements reduced but did not eliminate open access lands in the region. People could not reside within the confines of the park, nor could they settle in the tropical valleys where Shuar and earlier mestizo migrants had established homesteads. What did remain open for appropriation by settlers and investors throughout the 1980s were the lands in two extensive, auxiliary mountain ranges, the Cordillera de Kutukú and the Cordillera del Cóndor, to the east of the settled valleys and the Andean highlands. The Cordillera del Kutukú forms the mountainous backdrop to a colonist settlement and church in figure 3.2.

Beginning in the early 1990s, Shuar leaders took action to prevent large-scale land clearing in the auxiliary mountain ranges. They lobbied for the creation of a protected area in the Cordillera del Kutukú. The earlier struggles between the Shuar and the mestizos over landownership in the valleys had left a legacy that shaped later Shuar initiatives. The Shuar had organized early, during the 1960s, to prevent further losses of land to mestizo colonists. They created an umbrella organization, the Shuar Federation. Working initially with missionaries, Shuar leaders secured collective titles to the lands around each Shuar village in the valleys. The mestizos managed to gain control of lands that bordered the valleys at slightly higher elevations. The Shuar lost hunting grounds when the mestizos cleared these forests and planted pastures. The mixed outcomes from the Indigenous struggle to obtain land titles had a secondary political effect. It strengthened the Shuar umbrella organization as well as the local associations of Shuar villages.

FIGURE 3.2 Photo of Macas and the Kutukú mountain range in the background, 1969. Photo by Gary Hill.

While Shuar political organizations grew stronger, the economic conditions of Shuar households deteriorated. An inability to access credit from banks made it difficult for the Shuar to expand the size of their small cattle herds. Without savings or well-known sources of wage labor, the Shuar hesitated to move to cities with their cash-based economies. The young stayed in Shuar villages and subdivided their parents' landholdings when they died. Under these circumstances, the average size of Shuar landholdings in the valleys dropped precipitously, from sixty-six hectares to fourteen hectares over a twenty-five-year period, from 1986 until 2011.

The bleak economic prospects faced by the young Shuar encouraged them to move into the sparsely populated auxiliary mountain ranges and establish small farms in the narrow river valleys that run through the mountains. Given the rugged terrain and an absence of capital, most Shuar settlers chose to create small homesteads with gardens. They left the surrounding forests largely intact. To earn income, Shuar householders chopped down small numbers of commercially valuable trees, sawed them into boards, and hauled them out to the nearest road on the backs of mules. Shuar used the same mule trails to market the occasional cow or pig that they had raised on their homesteads.

To prevent incursions into the Kutukú mountains by mestizo colonists or deep-pocketed overseas investors, leaders of the Shuar umbrella organization pushed during the early 1990s for the creation of a protected forest in the mountains where Shuar households could live as inholders. The state approved the Shuar application for a protected area almost immediately, but for fifteen years it did little to implement the plan for a park. At that point, in 2008, the European Union and several international NGOs, working in concert with the Shuar Federation, amassed the funds necessary to create an acceptable and enforceable land use plan for the Kutukú protected area. The new plan created three zones of land use, with one zone of absolute protection covering 80 percent of the protected forest. The plan also included small zones of human settlement that accommodated Shuar households pursuing subsistence activities.

At the same time, in another, adjacent auxiliary mountain range to the south, the Cordillera del Cóndor, an association of Shuar villages enrolled a large, forested area of their lands in Socio-Bosque, a national program of payments for environmental services. The relatively modest payments from Socio-Bosque for environmental services appealed to the Shuar migrants because it allowed households to engage in culturally familiar daily activities such as gardening and receive payments for environmental services in a circumstance in which the rugged terrain prohibited alternative, more lucrative, but forest-destroying economic activities.[3]

This pattern applies more generally throughout Ecuador. Over decades of long struggles, Indigenous peoples in Ecuador gradually acquired state approved global titles to land. In contrast, mestizos, the other large class of smallholders in Ecuador, did not acquire secure titles to their land. This difference in land titles between mestizos and Indigenous peoples shaped their participation in REDD+ programs. Indigenous groups with global titles to their land qualified for participation in Ecuador's REDD+ program, Socio-Bosque, while mestizo smallholders, without secure titles, did not qualify. As a result, the Amazonian lowlands, with appreciable populations of Amerindians, contained 80 percent of the forested land enrolled in Ecuador's REDD+ program. The best statistical predictor of variations in REDD+ payments across Ecuadorian provinces in 2012 was the proportion of self-identified Indigenous people in a province's population in 2010. Provinces with large Indigenous populations received more payments for environmental services through the Socio-Bosque program.

The history of protected area growth on the eastern slope of the Ecuadorian Andes underscores the crucial role that the threat of further deforestation played in the creation of protected areas. The specter of further forest destruction in corridors along new roads, coupled with new colonist settlements, spurred the initial campaigns for conservation, beginning in the 1970s. Later, after the millennium, the Shuar, with a well-earned reputation for defending their own interests, built an alliance with deep-pocked conservation organizations to secure additional protected areas. By 2010, the Shuar had constructed a countercoalition that worked, particularly in more recent years, to protect large areas of forest from further destruction.

This history underscores the importance of alliances between Indigenous peoples, the state, and the conservation community in pressing for forest protection. The alliance exhibited several common features of corporatist processes: diverse actors contributing in different ways to a compact that secured a large, intact forest. The alliance has not been tested, at least in the initial years after its creation, perhaps because the mountains did not offer alternative economic opportunities and few people lived within or around the new protected areas.

General Patterns: The Continuing Loss of Tropical Forests in the Global South

While the rates of deforestation in southeastern Ecuador probably peaked during the 1970s and declined thereafter, forest destruction followed different trajectories elsewhere in the tropics. In Melanesia, for example, new road building projects triggered large increases in deforestation after 2010.[4] These long-term trends obscured short-term fluctuations in deforestation rates. In the Amazon basin, a dramatic decline in deforestation between 2005 and 2010 preceded a later acceleration in deforestation rates after 2015. Overall, between 1990 and 2019, the global extent of tropical moist forests declined by 17 percent. A further 10 percent of the remaining tropical forests were visibly degraded over the same time period, either through selective logging or fire. Many large operators used fire as a preliminary way to clear land, so fire-degraded forests during one period often became deforested lands during subsequent periods.[5]

Several patterns of change, consistent with the Great Acceleration in economic activities, have characterized the losses of tropical forests. The losses in Southeast Asia during the 2000–2010 decade came in larger blocks of cleared land than they had in previous decades. The growth in the size of clearings signaled a change in the identity of the people clearing the land. Smallholders, participating in agrarian reforms and clearing relatively small amounts of forest, may have driven deforestation during the 1960s and 1970s. By the turn of the century, plantations and companies that produced on a large scale for multitudes of consumers in distant cities had assumed a growing role in the destruction of tropical forests in Southeast Asia and South America.[6]

While deforestation did decline from the 2000–2010 decade to the 2010–2020 decade in both Latin America and Asia, it increased sharply in sub-Saharan Africa.[7] For the previous thirty years, deforestation had occurred more slowly in Africa than in the Americas or Southeast Asia, in part because the chief agents of deforestation in Africa were shifting cultivators who farmed small plots of land for household subsistence. Spurred by urbanization and continued rapid population growth, agricultural expansion in sub-Saharan Africa accelerated after 2000. Smallholders began to produce staple crops such as cassava for sale to urban residents. The rapid growth of urban populations also increased demand for fuelwood and charcoal which in turn accelerated the harvesting of trees in the semi-arid open woodlands around many cities. Concentric zones of deforested, denuded landscapes emerged around large cities such as Lusaka, Zambia.[8] In a departure from the patterns in South America and Southeast Asia, the expansion of export agriculture in sub-Saharan Africa did not account for a large proportion of the forest losses.[9] Smallholders cleared most forests to plant staple crops such as bananas and cassava that they then sold to growing numbers of rural and urban consumers. In sum, it seems that growth in consumer demand in domestic rather than international markets has driven recent increases in African deforestation.

The slowing rates of deforestation in the Americas and Asia, coupled with accelerating rates in sub-Saharan Africa, led overall to a modest global decline in deforestation rates after the millennium.[10] This decline, while visible, should not obscure the immense losses in biodiversity and the large scale of greenhouse gas emissions that accompanied the continuing deforestation in the tropics between 2010 and 2020. Figure 3.1, a photograph of recently cleared old growth forest in the Ecuadorian Amazon, provides graphic evidence of the ecological destruction that occurs when farmers clear land of old growth forests. The continued destruction of these forests links directly to the accelerated growth of the human enterprise in the Global South as well as in the Global North.

"Avoided Deforestation": The Resistance to the Continuing Loss of Forests

Since the 1970s, activists drawn from the natural sciences, from Indigenous groups, and from conservation groups have tried to prevent farmers, ranchers, and loggers from destroying or degrading tropical forests. When successful, these efforts leave the forests standing. In this sense, protected tracts of land have avoided deforestation. As the continuing deforestation in the tropics suggests, forest preservation has been more the exception than the rule in the outcomes of conflicts over the use of forested land. At best, conservation efforts have slowed tropical deforestation.

Conservation Advocates for Protected Areas

Since the 1970s, natural scientists and Indigenous peoples, have persistently opposed the destruction of forests. Natural scientists voiced concerns about the losses of biodiversity that occurred when farmers converted old growth forests into pastures for cattle or fields for cereal crops. Initially, natural scientists worked separately from Indigenous groups in campaigns to prevent deforestation.[11] After 2000, environmentalists, organized into NGOs, and Indigenous peoples, organized into political pressure groups, have joined forces. Their joint efforts at resistance have rarely stopped the construction of roads or the expansion of agriculture into new areas, but they have succeeded in having extensive tracts of forested land set aside as protected areas or as reserves for forest-dwelling Indigenous peoples.

The initial campaigns for conservation during the 1970s, 1980s, and 1990s coincided fortuitously with international state building pressures in which influential ministers in the governments of the Global South concluded in an isomorphic dynamic that states in the developing world should, like states in the developed world, have national park systems. To this end, Brazil's head of National Parks, Maria Padua, convened in the mid-1970s a panel of international scientists with conservation interests to map out a 20,000 km^2 expansion in Brazil's system of national parks. Following the recommendations of the scientists, Padua located many of the new parks in the Amazon basin.[12] The Brazilian example inspired officials in Ecuador to do likewise, which in turn contributed to the creation of the Sangay National Park in the mid-1970s.

Once created, these parks could, in moments of crisis, be augmented. In 1998, after a several year hiatus in reporting rates of deforestation in the

Amazon basin, Brazil reported particularly high rates. This report spurred alarmed comment by scientists and environmental activists in Europe and North America. The political pressure appeared to produce results, as a month later, the Brazilian government announced a large expansion in its national system of protected areas. It created a particularly large number of parks in the sparsely populated rainforest in the northwestern quarter of the Amazon basin, in the province of Amazonas.[13] In this instance and in the Ecuadorian instance described earlier, the sparse populations in the regions designated for parks appealed to government administrators because the small numbers of inhabitants in the newly designated parks promised to limit political resistance to the creation of the parks.

While the pace of park creation has declined in Latin America from the first to the second decade of the twenty-first century, governments in the western reaches of the Amazon basin have continued to create parks in sparsely settled zones.[14] The Chiribiquete park (2,800,000 hectares) created in 2013 in the Colombian Amazon and the Sierra del Divisor park (1,300,000 hectares) created in 2015 in the Peruvian Amazon exemplify this trend.[15] Indigenous peoples played important roles in the campaigns that led to the creation of both parks.

Political conflict fueled the push for smaller scale parks in many places. This dynamic typically begins when investors propose new economic developments. The proposals trigger conflicts between development interests and conservationists. The conflicts usually end with resource partitioning. In an attempt to please everybody, governments officials subdivide the disputed lands. Developers destroy forests in one tract. To mollify the opponents of development, government officials create a new protected area in the other tract of land. While new, these protected areas are also quite small. After all, they only represent a portion of the lands in dispute.

This resource-partitioning dynamic has occurred in a wide range of settings. For example, in 2001, the central government of Ecuador planned to build a second pipeline from its Amazonian oil fields over the Andes to a port on the Pacific coast. The proposed route of the new pipeline went through Mindo, a subtropical valley favored by birders for its rich avian biodiversity. Predictably, the birders, with support from local residents and international environmental NGOs, objected to the pipeline's route. The pipeline's opponents demonstrated in the capital, blockaded the pipeline construction site, and took legal action to stop the pipeline's construction. Their efforts did not succeed in stopping the pipeline, but in an attempt to defuse the political conflict, government officials did enlarge the wildlife reserves around Mindo.[16]

While the size of these new protected areas has been small, the large number of new preserves has made the total area protected large. In New England one-half of all protected forests have been conserved in this fashion since the 1990s.[17] In northern New Jersey 30 percent of all land achieved protected status through this dynamic between 1970 and 2005.[18] These efforts at conservation in contiguous areas laid the foundation for regional conservation partnerships in which NGO personnel, government officials, and landowners came together repeatedly in corporatist efforts to enlarge protected forest areas.

Similar sequences of events happened in very different contexts. In an exurban community west of New York City, American Telephone and Telegraph (ATT) purchased three hundred acres of undeveloped, forested land and announced plans to construct a training facility on the land. The neighboring landowners, concerned about the loss of a "country atmosphere," contested the proposal in hearings and later in the courts. After a considerable delay caused by the conflicts, the neighbors, with both public funds and support from both local and national environmental NGOs, purchased a large portion of the original three hundred acres and turned it into a nature preserve. Developers revised their original plan and built single-family homes on the remaining tracts of land.[19] Here, as in the case of Mindo, Ecuador, continued real estate development in a context of widespread environmental concern prompted additional resource partitioning and the creation of small, new protected areas in the immediate vicinity of the new development.

Designation of a primary forest as a protected area does not insure the protection of a particular forest, of course. Settlers have cleared forests inside park boundaries throughout the Global South, but it is also true that the likelihood of deforestation declines when government officials establish protected areas in a place.[20] With this qualification, it is still fair to conclude that incremental gains in this type of resistance to deforestation continue to occur. With large pledges of conservation dollars from ultra-wealthy individuals such as Yvon Chouinard and Jeff Bezos, funding for the purchase of protected areas should continue to become available.

Some forests, in particular national forests, have received half-measures of protection. Brazil and the United States have established the most extensive systems of national forests. State foresters manage these forests for multiple uses, including clear cuts of mature trees. The frequent harvesting of these trees prevents these forests from becoming carbon sinks. In the first two decades of the twenty-first century, environmental groups in the United States lobbied to restore these carbon sinks by ending logging in the national forests.

The accumulating concerns with climate change and biodiversity losses have begun to alter local attitudes toward protected areas, especially in regions where the growth of ecotourism has had a significant economic impact. In countries such as Costa Rica and Ecuador with large streams of ecotourists, fewer residents subscribe to the idea that biodiversity conservation limits local livelihoods. Rural peoples living adjacent to parks have begun to see protected areas as the source of new revenue streams.[21] Under these circumstances, local support for new ecotourist destinations have made it somewhat easier to establish or enlarge protected areas.

Indigenous Peoples and the Creation of Reserves

Indigenous peoples whose livelihoods depend on non-timber forest products have become frequent defenders of forests since the 1970s as they acquired control over more land. During the last three decades of the twentieth century, Indigenous groups in the Global South secured control over 250 million hectares of land, an area the size of Argentina.[22] Much of the Indigenous-designated land was forested. With Indigenous peoples pursuing modest, forest-based livelihoods, the extension of Indigenous control to these lands made their preservation as forests much more likely.

Indigenous groups such as the Shuar now have legitimate claims to more than one-quarter of the earth's land area and approximately 40 percent of the world's protected areas. The domain of Indigenous peoples contains a high proportion (67 percent) of lands in a natural state. The undeveloped condition of these lands has increased the likelihood that they would become the targets for campaigns to protect these lands.[23] Once Indigenous peoples have acquired control over areas, they have proven to be just as effective in preventing the development of these lands as the state bureaucracies charged with protecting park lands.[24]

The preservation of forests in Indigenous reserves has frequently hinged on the continued acceptance of forest-based livelihoods by Indigenous peoples. These livelihoods usually have deep cultural roots, but they also have an underlying economic logic that begins with the poverty of Indigenous peoples living in remote settings. The absence of easy transit for people and for goods reduces the economic opportunities available to Indigenous people in these regions. The absence of economic opportunities impoverishes Indigenous peoples, but it also reduces the likelihood that they would cut down the forests in the reserves.

The same dynamic would make arrangements for sustainable development more palatable to Indigenous households living in forested preserves. Small payments for the environmental services entailed in preserving a forest might, for example, prove attractive to Indigenous households who do not have lucrative alternative livelihoods that would entail destroying primary forests. Coalitions between Indigenous peoples and large international NGOs, stemming from their shared preferences for sustainable development, have, as in the Ecuadorian example described earlier, helped Indigenous communities accumulate the political power necessary to establish the reserves.

This pattern of Indigenous-supported forest preservation has occurred across a wide range of places. Like the Shuar in Ecuador, Indigenous people in Indonesia, in alliance with activists from international environmental NGOs, mounted effective opposition to proposals for large-scale oil palm plantations in Kalimantan, the Indonesian portion of the island of Borneo.[25] Indigenous groups have also been vital participants in upland tree planting projects in interior Southeast Asia.[26]

Political contingencies sometimes mark the creation of the reserves. Seesaw political struggles occur in which Indigenous organizations best rival claimants at one time, only to see their gains reversed at a later date. Reversals of protection have occurred when small Indigenous groups, without political allies, face powerful agricultural interests, as occurred in Rondônia in Brazil in 2021 when the state legislature reduced the size of the Jaci-Parana reserve by 90 percent. The reserve was established during the 1990s for rubber tappers whose numbers have since declined as cattle ranching expanded in the region.[27]

Overall, undesignated forest lands continue to diminish, with new Indigenous preserves continuing to appear at politically opportune moments. For example, Brazil created two new Indigenous preserves, the Turubaxi-Tea Indigenous territory in 2017 and the Kaxuyana-Tunayana Indigenous territory in 2018, right before a conservative regime, opposed to Indigenous claims, assumed power in Brazil.[28] The creation of transnational Indigenous NGOs such as Nia Tero has facilitated the application of political pressure by Indigenous groups over governance issues. Here, too, as with protected areas, incremental gains continue to occur in the struggle to avoid deforestation.

The increases in the extent of Indigenous preserves, as in the Ecuadorian example, while impressive, still leave large areas vulnerable to destruction. The gains in Indigenous preserves, coupled with growth in the extent of protected areas, have established protections over two-thirds of the Brazilian Amazon. As much as one-third of the Amazon forest remained without protections in 2022. The recent losses of Brazilian forests have concentrated in these unprotected areas.[29]

Community Forests: Villagers in Search of Sustainable Development

In large, densely populated rural regions in Latin America and South Asia, but not in sub-Saharan Africa, villagers in long established communities have become the owners of common property forests close to their homes.[30] These forests, while not subject to as strict a level of protection as most parks, do enjoy a significant measure of protection in many places. In

Mexico, more than in other places, historical circumstances spurred the creation and expansion of community forests that now sequester carbon and provide livelihoods for heretofore impoverished small farmers. These villagers created the community owned forests in *comunidades* stemming from the colonial period and on *ejidos*, collective land grants made to landless farmers beginning in the 1930s. The community forests make up 60 percent of all forests in Mexico. They averaged 3,793 hectares in size in 2015.[31] These pine and pine–oak forests existed mostly in a naturally regenerating belt of temperate forests in Mexico's mountainous regions.

Since the 1970s, *ejidatarios* (residents of *ejidos*) have gradually constructed wood product enterprises around their forests. They purchased machinery for harvesting, constructed sawmills, and started wood product industries in the villages. Public officials lent a hand in these endeavors. They ratified the villagers' land tenure in the early 1990s. They provided expertise and subsidies that enabled the villagers to get their forest-based enterprises up and running. In this sense, the Mexican state provided vital assistance that enabled the creation of common property enterprises in the village forests. The villagers who run these forest enterprises seem more focused on supplying jobs and services to village residents than they are in maximizing profits from the enterprises.[32]

Deforestation, albeit at slowing rates, has characterized Mexico's forests since the 1970s. Against this somber backdrop, the community forests have stood out because, during this period, they registered small gains in forest cover. The growing value of the pine and pine-oak forests seemed to have contributed to their growth in extent. When the *ejidatarios* did harvest trees, they took fewer trees per hectare than did harvesters elsewhere, presumably because the lower harvest rates provided a sufficient income for village residents and promoted the long-term sustainability of the forest-based enterprises in the villages.[33] The lower rates of harvest in Mexico also made it more likely that these common property forests would serve as net carbon sinks.

The community forests that provided the most environmental services were middle-sized. They seemed able to both deliver considerable environmental services and earn the *ejidatarios* an adequate income. Sustainable development in these places stemmed in part from the ample base of human resources that village leaders could draw upon to do the work of timber extraction and forest restoration. These circumstances suggest that dense rural populations in regions such as South Asia and Central America represented a crucial enabling condition for community forests that could both sequester carbon and secure livelihoods for their owners.[34]

The enabling conditions for the sustainable development of community forests in Mexico have been both political and economic. The Mexican Revolution in the early twentieth century provided the political impetus for widespread agrarian reforms that made the collective ownership of land in *ejidos* and *comunidades* available to large segments of the rural poor. For the remainder of the twentieth century, public officials sought to further the redistributive impetus of the revolution through a series of laws and programs that enabled forestry in the forest rich regions of Mexico. The *ejidatarios* took advantage of these opportunities, beginning in the 1960s. Upward trends in prices for Mexico's prevalent pine trees provided crucial supporting opportunities for the community foresters. Heretofore unfamiliar groups of people, members of *ejidos*, public officials, and industrial foresters, came together and learned new skills that enabled the community forests to flourish as economic enterprises. Beginning in 2003, Mexico increased these cooperative efforts by establishing a national payments for environmental services (PES) program that successfully promoted forest cover conservation through payments from the national government.[35] From year to year, the program discussions, agreements, and corresponding conservation payments resemble corporatist processes.

Payments for Environmental Services: Can Assistance from a Wealthy Country Stop Deforestation?

Most analysts of tropical deforestation, educated in market-based societies, bring a perspective steeped in market-based solutions to their work. Frequently, they self-identify as economists. To do their work, they deploy the analytic tools of microeconomics, the most powerful paradigm in the social sciences. Given these circumstances, it is perhaps not surprising that PES, a market-based mechanism, attracted a great deal of attention as a possible solution to the persistent problem of deforestation when observers first began talking about it two decades ago.

REDD+ initiatives have usually occurred in discrete geographical juris-dictions such as countries or provinces. Some countries, like Ecuador, estab-lished their own programs, where the central government provided the funds for carbon sequestration payments to forest owners. Many REDD+ programs took the familiar form of development assistance, with funds flowing from wealthy countries such as Norway to organizations in par-ticularly poor, forest-rich places. REDD+ had become "aidified." A multi-tiered organizational architecture characterized many of these projects, with funds flowing from international donors to government ministries and then to individuals and Indigenous community councils. To acceler-ate the provision of environmental services through these organizational links, Norwegian officials embarked in 2008 on the world's largest bilat-eral, REDD+ assistance program. They contracted to pay for environmen-tal services in four countries: Brazil, Guyana, Indonesia, and Tanzania.

TANZANIA

Tanzania and Norway made the first of these bilateral agreements. Much of the substance of the agreement was drafted by the staff at Norway's embassy in Tanzania. It entailed a set of nine small contracts with NGOs that focused on carrying out forest friendly activities in villages. The NGO activities did not entail the results-based compensation for individ-ual landowners that constitutes the core of the REDD+ idea.[36] The implicit focus on village level assistance in this agreement may reflect the histori-cally strong political tradition of governance through village elders estab-lished during the Ujamaa period in Tanzania. The decentralization of forest governance across sub-Saharan Africa after 1990 probably reinforced this tradition of local governance.[37] Donor surveillance of expenditures did not produce visible frictions in the administration of the bilateral agreement.

BRAZIL

Norwegian officials wanted to commit a large sum of money to the preservation of the Amazon forest because of its symbolic value as the world's largest tropical rainforest. Ideological compatibility in 2010 between Norwegian officials and center-left Brazilian officials led to a quick agreement through which the Norwegians committed one billion dollars to be spent on carbon sequestration projects identified by the Amazon Fund, a "to be created" Brazilian agency. The need to create a recipient agency slowed down the disbursement of funds, but it did begin. Unfortunately, Jair Bolsonaro, as soon as he took power in 2019, abolished the Amazon Fund. With Bolsonaro's defeat in 2022, his successor, Lula da Silva, and Norwegian officials have indicated a willingness to revive the original program. The original billion-dollar Norwegian commitment, while very large for a bilateral assistance program, did not come close to compensating Brazil for all of the additional carbon sequestration that occurred through the slowdown in deforestation during the ten years before the agreement between the two countries.

GUYANA

The Norwegian officials' impulse to spread the wealth led to the negotiation of a bilateral agreement between Guyana and Norway. Guyana, like Gabon and Suriname, has large forests and low deforestation rates. In these settings, small year-to-year fluctuations in forest cover do not generate calculable gains in environmental services compared with the large scale and calculable changes in greenhouse gas emissions that occur in countries such as Indonesia with historically high deforestation rates. Norwegian officials wanted to figure out how to incentivize forest conservation in low deforestation countries. To this end, they authorized a series of payments to Guyana, a country with little deforestation. The payments from Norway to Guyana went to sometimes questionable ends such as the construction of a dam, but the disbursements of funds did occur, so the contract fulfilled its role as a pilot project.

INDONESIA

The bilateral agreement between Norway and Indonesia has had the most problematic history of all of the agreements. It was the last of the agreements to be signed, in May 2010. The discussions that led up to the signing of the letter of intent featured contentious interchanges about Indonesian forest policies that had contributed during the 2000–2010 decade to the world's highest volume of greenhouse gas emissions from deforestation. The exploitation of extensive peat lands in Kalimantan contributed to Indonesia's high levels of greenhouse gas emissions. These discussions included an Indonesian commitment to a moratorium on new logging concessions in primary forests. The moratorium did not apply to forests that had already been degraded by logging and land clearing, a provision that disappointed environmentalists. The high visibility of Indonesian deforestation persuaded the Norwegians to commit to another billion-dollar contract, equal in size to the Norwegian contract with Brazil.[38] The execution of the new policies, the disbursement of some funds, and discussions about further disbursements continued until 2021 when Indonesian leaders decided to drop out of the bilateral agreement owing to dissatisfaction with the rate of distribution of the Norwegian funds.[39]

Comparisons across the four contracts underscores the importance of political variables in the ability of the governments to carry out these bilateral agreements. Lula da Silva's first government in Brazil had such a sterling reputation among environmentalists that donors such as Norway were willing to cede control over streams of REDD+ income to Brazil's Amazon Fund. A quite different but seemingly workable dynamic emerged in Tanzania, where donors ceded control to locally oriented nongovernmental organizations (NGOs) out of an assumption that the combination of local leaders and NGO personnel would arrive at an acceptable way of disbursing the funds. A very different political dynamic, founded on mistrust, prevented the full implementation of the Indonesia–Norway pact. These patterns across the different bilateral pacts underscore the pivotal role that politics played in assembling coalitions capable of implementing REDD+ programs.

The "aidification" of REDD+, its transformation from a market-based conservation tool into a product of bilateral assistance programs, probably accounts for the growing salience of politics in the implementation of the program. In its initial conceptualization, REDD+ had anonymous actors in a market delivering streams of money to landowners in ephemeral market transactions. The adopted format of "aid for reform" divided donors and recipients along political lines between countries and called for very different kinds of implementing activities from politicians in the donor and recipient governments. The differences in levels of affluence between countries predisposed the negotiators to concerns about very different issues, political support for overseas assistance in Norway versus the willingness of local power brokers in rural areas of Indonesia to conserve tropical rainforests. In some instances, as in Tanzania and Brazil, recent historical experiences made it possible for the two groups of REDD+ politicians to bridge their differences. In other instances, as in Indonesia, the agreement collapsed under the weight of political differences.

The overall record of the Norwegian bilateral effort suggests substantial success in initial efforts at implementing REDD+. Three of the four bilateral assistance programs achieved a modicum of success even though REDD+, in its original market-based conception, never began. Going forward, REDD+ could proceed along two tracts. One line of institutional development would proceed as a voluntary, market-based activity with brokers between affluent buyers of carbon offsets and landowning sellers. The other line of development would emerge in places characterized by market failure where buyers and sellers could not be found. In these instances, government officials would step in and negotiate compacts, in a corporatist process, between aid programs and carbon sequestering landowners. National payments for environmental services programs in Ecuador and Vietnam approximate this situation.

Leaders: Can "Command and Control" Deter Deforestation?

Under some circumstances, lobbying elites may prove to be the most efficacious way to preserve rainforests. Brazil's history of protected area and Indigenous preserve expansion, in the late 1970s and again in the 1990s, demonstrates the far-reaching impact of leaders on the preservation of forests. The history of changes in the Brazilian forest code leads to the same conclusion. Brazil's leaders adopted a forest code during the early 1930s that required landowners to maintain 50 percent of their land in forest. Legislators insisted on forest reserves in order to retain adequate supplies of fuelwood in communities. In 2001 after an extended period of rising international concern over increasing rates of tropical deforestation, Fernando Henrique Cardoso, then president of Brazil, increased the reserve requirement to 80 percent for landowners in the Amazon basin.[40] More frequent and higher resolution satellite imagery, coupled with clarified boundaries between properties, made it possible for authorities to ascertain when recent land clearing by a landowner had violated the 80 percent reserve requirement. In some instances, forest rangers made high-profile visits to the farms where the violations occurred. When the violations clustered in the same municipalities, the agrarian bank imposed credit restrictions on farmers from those municipalities.[41]

Two patterns have shaped this elite-influence approach to preserving forests. First, surveys display dramatic differences from large to small properties in the extent of forests. In a survey on the fringes of the Brazilian Amazon, properties under 150 hectares had less than 21 percent of their land in forests. In contrast properties in excess of 2,500 hectares had more than 62 percent of their lands in forest. As these figures suggest, compliance by Brazilian landowners with the 80 percent reserve requirement was low.[42] In a move that reflected discontent with this mandate after five years of substantial declines in deforestation, the Brazilian legislature revised the Amazon reserve requirement in 2012, lowering it to 50 percent.

Second, elite and NGO-driven efforts to protect the Brazilian forests have exhibited political momentum. Cardoso's 80 percent mandate in 2001 encouraged NGOs to press for additional restrictions. They campaigned for restrictions on the purchase or sale of commodities grown on recently deforested lands. In 2006 international soybean traders and soybean processors in Brazil, under pressure from Greenpeace, agreed not to purchase soybeans that farmers had grown on lands deforested during the previous year. Similarly, in 2009, prominent meat packing companies agreed, again under pressure from NGOs, not to purchase cattle raised on recently deforested pastures in the Amazon basin. European importers of Brazilian beef reinforced these restrictions by placing a moratorium on the purchase of Brazilian cattle raised on recently deforested pastures.[43] After the Brazilian initiatives, legislators in the contiguous Latin American countries of Paraguay and Argentina considered and in some instances passed zoning laws or moratoria designed to reduce or stop deforestation in the dry forests of the adjoining Chaco region.[44]

These initiatives by leaders, states, and NGOs had visible effects. Pasture expansion slowed down in the Brazilian Amazon between 2004 and 2010. Importers from overseas did increase the flow of soy imports from regions with more restrictive regulations over soy-related deforestation. Other studies suggest that leader-led political controls had few effects. Larger scale analyses that included Argentina, Paraguay, Bolivia, and other regions of Brazil did not show a discernible slowdown in deforestation rates after the adoption of more restrictive regulations over land clearing. The imposition of more restrictive land clearing regulations in Brazil in the 1990s spurred a movement by large Brazilian growers to places such as the Chaco region of neighboring Paraguay. In effect, new rules, enacted by leaders in one country, created "deforestation havens" in adjacent countries and spurred the creation of commodity frontiers and deforestation in those places.[45] In these instances, the activities of the deforesting growth coalitions leaked into adjacent jurisdictions. As intended, the new rules reduced deforestation in one jurisdiction, but at the cost of increasing it in other, adjacent jurisdictions. Leaders proclaimed changes, and legislators enacted them, but, if government officials did not enforce the new rules, little change occurred. In some settings, such as Argentina, lax enforcement diminished the protective effects of the new regulations on forests.[46]

Conservation mandates also proved ineffective when they had to contend with countervailing trends among consumers. In Brazil, growth in domestic demand for beef and soybeans diminished the overall effects of the new restrictions on forest cutting. Small meat packing plants and soybean processors provided beef and soy products to the Brazilian markets. These organizations usually did not comply with the new deforestation laws in their purchase decisions. Ranchers and farmers who violated the deforestation restrictions could sell their cattle and soybeans through these companies. Under these circumstances, growth in domestic demand for beef and soy products supported continued deforestation in the Brazilian Amazon.[47]

The rates of deforestation rebounded with the ascent of the conservative Jair Bolsonaro to the presidency of Brazil in 2019. Ranching and logging interests saw the ascension of a sympathetic right-wing political leader to power as an opportunity to destroy forests without fear of arrest or prosecution. A further change in political prospects, with a prospective transfer of power away from a probusiness leader like Bolsonaro to an environmentalist like Lula da Silva, appeared to spur further deforestation in 2022 before the transfer of power.

Five years after Brazilian politicians established their moratoria on beef and soybean induced deforestation, Indonesian politicians adopted a similar approach to deforestation. In 2010, they declared a moratorium, to begin in 2011, on the granting of additional concessions of primary forest for logging and palm oil production. Indonesian leaders extended this moratorium three times after 2011 and made it into a permanent prohibition in 2019. Concessions granted before 2011 continued to operate. Municipal leaders launched sometimes successful campaigns to extend the boundaries of concessions during election years.[48] The exemptions of established concessions and secondary forests from the moratoria on cutting enabled continued deforestation. So did the changes in the boundaries of existing concessions. These exceptions have made the overall effects of the moratoria on Indonesian deforestation difficult to determine. The Brazilian example has played a difficult to determine role in the policies adopted by Indonesian leaders to combat deforestation. Again, market forces may have shaped the overall trends in deforestation in Indonesia. Oil palm prices peaked on the world market in 2012, and Indonesia saw its highest deforestation rates in 2014, three years after the state imposed the moratoria on concessions. Since then, Indonesian rates of deforestation have declined from these highs.[49]

It is not an accident that Brazil and Indonesia, the two most prominent countries in terms of the extent of recent tropical forest losses, have taken the lead in creating rules to prevent deforestation and in coercing compliance to these rules. They have had a mixed record of success, on again and off again enforcement in Brazil and crucial exceptions to the rules in

Indonesia. Enactment and effective enforcement of rules against deforestation seems to require strong states that have the capacity to mobilize large sectors of their populace to pursue a common goal. Put differently, rules-based protection of forests requires an institutional strength that many states do not currently have. The fostering of forest conservation and restoration through corporatist political processes does not seem feasible when states are weak.

The focus on state capacity underscores the role that elites have played in reducing deforestation. Their effect on rates of deforestation has been most evident in Brazil where two presidents, Fernando Henrique Cardoso and Luiz Inácio Lula da Silva, spurred the implementation of effective anti-deforestation policies. The retrogressive impact of Bolsonaro and upward spike in the deforestation rates in the Amazon provide grim testimony about the sometimes nefarious impacts of elites on the destruction of forests. Some conservation policies are not vulnerable to leader-led assaults. Because ecocertification programs depend on the willingness of distant consumers to pay premium prices for certified products, they would not be vulnerable to leader-led political pressures to increase the destruction of rainforests.[50]

Avoided Deforestation in the Larger Context of Climate Stabilization Efforts

The presentation of the five ways to avoid further losses of primary forests misconstrues these efforts in one respect. The text presents them as discrete alternatives when, in practice, they often work in combination. Perhaps the clearest example of this combinatorial pattern involves Indigenous preserves and REDD+ projects. Indigenous groups have become active participants in REDD+ projects in a number of countries. The Ecuadorian example with the Shuar provides a particularly clear example of this pattern. Similar interests have brought conservationists, grouped together in environmental NGOs, together with Indigenous groups and sympathetic government officials. In this sense, the coalitions involving the Shuar in the rainforests of southeastern Ecuador seem representative of forest-saving coalitions elsewhere in the world. Given the prevalence of these coalitions, the resistance to rainforest destruction, while always in a reactive stance, has more political strength than is sometimes acknowledged. The defenders of the forests can and do coalesce in opposition to plans for the further destruction of rainforests.

The avoided deforestation efforts described earlier began as an effort to conserve biodiversity, but the more comprehensive climate stabilization movement inaugurated in the 2010s has largely subsumed the movement to conserve biodiversity. With the two movements cooperating and coordinating more, their goals have converged. They both emphasize the preservation of primary forests. The climate stabilization groups to a greater extent and the biodiversity groups to a lesser extent also focus on the restoration of forests on degraded agricultural lands.

The importance of these forest expansion and restoration efforts have grown as the overall extent of deforestation and the feedback effects of deforestation on carbon sinks in the tropics have become more evident. Recent research on carbon emissions from the different regional landscapes in the Brazilian Amazon has established that forested landscapes lose their capacity as carbon sinks when cleared land increases. In the Southeast, the most deforested quadrant of the Amazon basin with approximately 30 percent of the land denuded, land clearing has changed the climate. It has lengthened the dry season. The increased dryness has accelerated the number of forest fires and associated greenhouse gas emissions. The emissions from the fires have in turn converted the southeastern Amazon landscape from a carbon sink into a carbon source.[51]

These findings about the feedback effects of deforestation on carbon fluxes have several important implications for climate change policy. First, they underline the continuing importance of avoiding further deforestation. A successful global effort to avoid deforestation would reduce global greenhouse gas emissions by 16 percent to 19 percent per annum. Second, efforts to reforest largely deforested tropical landscapes such as those in the southeastern Amazon basin would convert these landscapes from carbon sources to carbon sinks.[52] In quantitative terms, a robust, worldwide effort to foster regrowth in the tropics might reduce global greenhouse gas emissions by another 8 percent to 11 percent per year.[53] In other words, for forests to serve as a natural climate solution, societies must not only avoid further deforestation of old growth forests but also promote the regrowth of forests across landscapes where they do not currently exist. The next four chapters discuss the different forms that this reforestation would take in these predominantly agricultural landscapes.

4

• • • •

Rural–Urban Migration, Land Abandonment, and the Spread of Secondary Forests

Of the most common forms of forest expansion in the twenty-first century, agroforests, silvopastures, forest plantations, and secondary forests, the latter brings with it the most biological riches. While secondary forests do not contain the diversity of life forms found in old growth forests, they contain more biodiversity than forest plantations, agroforests, and silvopastures.[1] Secondary forests also typically sequester carbon from the atmosphere at higher rates than do most agroforests and plantation forests, especially if the secondary forests survive long enough to become old growth forests.[2] The biological richness of secondary forests has persuaded almost all ecologists that secondary forests should be the preferred form of forest restoration.[3]

Land abandonment prepares the ground for secondary forests, but it does so in distinct ways from region to region. Four case studies in the following pages of the land abandonment–secondary forest transition illustrate the variable ways in which secondary forests have expanded since the 1870s. The descriptions include the biogeographical setting for regrowth, the political-economic processes that precipitated land abandonment, and the accumulating pressures, if any, to chop down the emerging secondary forests. Considered together, the descriptions of the four cases should answer questions about the contexts and the immediate circumstances that have fueled the expansion of secondary forests. They outline two trajectories of secondary forest expansion, one through land abandonment by smallholders and the other through fallowed land by large landowners.

Forest Regrowth in a Town That Went Downhill:
Northeastern United States

In the first few decades of the nineteenth century, young European colonists from coastal settlements in the northeastern United States began moving inland and upstream into the Appalachian highlands in search of land to farm. They found depopulated places, inhabited by small Native American groups decimated by disease and warfare with European settlers. The colonists settled the less accentuated patches of terrain where they could practice rainfed agriculture, growing corn, potatoes, and turnips. They also tapped maple syrup from trees and grazed sheep and cattle on high-elevation fields, above three hundred meters in altitude. To clarify the boundaries of farms, clear fields of stones, and create holding pens for sheep, farmers piled stones in walls that enclosed fields. Figure 4.1 shows one of these stone walls in a now overgrown field in the New Jersey Highlands.

FIGURE 4.1 Photo of a stone wall on a forested hillside, northwestern New Jersey. Photo by the author.

The large numbers of intersecting stone walls in today's secondary forests testify in a muted way to the dense populations of poor farmers who earned meager livelihoods from these upland agricultural landscapes in nineteenth century New England. One late twentieth century survey found 380 kilometers of stone walls running through one largely forested, 18 km by 10 km, former farming community in western Massachusetts.[4] Roads, really wagon tracks, ran from population centers in the river bottoms up the narrow valleys of tributaries and onto the saddles between the mountains.

Except for the stone walls, the settlers made few, if any, capital improvements to these upland farms. The old growth forests provided wood for house construction and fuel for cooking and heating. The quantities of available manure were not sufficient to fertilize the fields, so harvests diminished in size with continued cultivation. Deteriorating market conditions added to the difficulties faced by the mountain farmers. The trade in agricultural commodities that sustained these farm families began to shrink during the 1820s when the completion of the Erie Canal linked the cities on the Atlantic Coast to the growing numbers of farms in the agriculturally rich Midwest. The upland New England farmers, working progressively more exhausted soils, could not match the quality and the price of the agricultural commodities produced by the Midwestern farmers.[5] For this reason, the New England farmers lost market share and a vital source of income. By the mid-nineteenth century, some of these farmers and their offspring, most notably their daughters, had decided to move elsewhere. Some of them did not go far. They moved from hillside farms down to growing manufacturing centers in the valleys where they could find work in newly constructed, water-powered textile mills.

Competitive pressures in the form of more remunerative wage labor in textile mills and more cheaply produced grains from more richly endowed agricultural regions had triggered 'agricultural adjustments'[6] in the uplands. By the 1850s, farmers had begun to abandon these lands. Figure 4.2 maps this process of land abandonment in one New Hampshire community, Lyme, on the Connecticut River. The land rises from 130 meters above sea level along the Connecticut River in the west to mountains that exceed 1000 meters above sea level in the east. The lighter dots denote abandoned

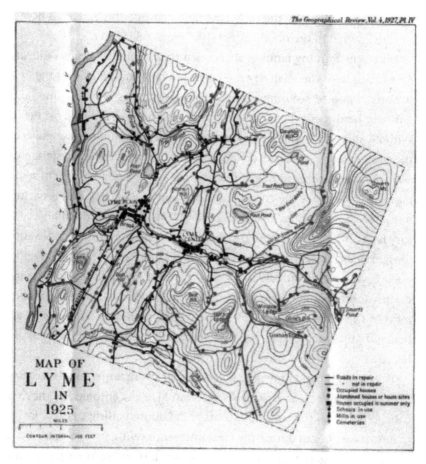

FIGURE 4.2 Map of land and road abandonment, Lyme, New Hampshire, 1927. Courtesy of *Geographical Review.*

farmhouses, and the lighter lines indicate abandoned roads. First, the families abandoned the farms, and once all the families along a road had left their farms, the township stopped maintaining the roads up to the farms. While abandoned farmhouses and roads could be found all over the town in the early twentieth century, they clustered in the eastern highlands where the climate shortened growing seasons and the remote location complicated agriculture. Smallholders abandoned the highest elevation and most remote farms first. Between 1860 and 1892, people abandoned one-third of the homes in the eastern reaches of Lyme. Literally, the town went downhill over time. By the early twentieth century, the inhabited home-steads clustered in the floodplain of the Connecticut River.[7]

When the farming families abandoned their farms, the forests quickly took their place. The relatively even distribution of rainfall across the four seasons promoted seedling survival, so the seedlings of white pines and northern hardwoods spread from the woodlots to the abandoned old fields. Without surveys of land cover, observers could do no more than guess at the exact extent of the early deforestation and subsequent reforestation in many towns. The magnitude of the land use conversion was large in some places. Farmers near the crossroads community of Petersham, Massachusetts, had cleared nearly 90 percent of their land by 1830. More than a century later, in 1960, Petersham had 86 percent of its lands in forest. Seventy to eighty percent of the cleared, agricultural lands in Petersham had reforested between 1850 and 1950.[8] This pattern of farmland abandonment occurred throughout the northern half of the Appalachian chain of mountains in the eastern United States. The regrowing forests sequestered carbon, most likely more than the two tons of carbon sequestered per year per hectare by old growth forests in the region.[9]

The gradual abandonment of upland Appalachian farms and the accompanying spread of secondary forests over the old fields continued for the first three-quarters of the twentieth century. A countervailing trend of losses in forest cover began during the mid-nineteenth century when prosperous city dwellers began buying up choice lakeside lots and picturesque mountain farms to use for vacations during the summer months.[10] This trend gathered force, to the point where, during the last quarter of the twentieth century, the losses of forest to real estate development exceeded the gains in forest cover on abandoned farms.[11] By the late twentieth century, few upland farms remained, so land abandonment declined. Meanwhile, improvements in the interstate transportation network spurred the construction of second homes in the mountainous regions closest to the coastal cities. Because the loss of forests to real estate developers occurred on a spot-by-spot basis, the extent of individual land clearings was not great, but they did accumulate and counteracted reforestation on the remaining mountain farms. In sum, the reforestation of the northern Appalachians occurred for a limited period of time. It emerged during the mid-nineteenth century, and it had disappeared by the late twentieth century.

Reforestation in a Fragmented Agricultural Landscape: Northern Portugal, 1960–1990

Like most farming communities in Western Europe, farmers in northern Portugal saw an expansion in forest cover during the first eight decades of the twentieth century. A large area in northern Portugal, extending across ninety-nine municipalities, saw forest cover increase from 24 percent to 32 percent of all land between 1906 and 1970.[12] Shortly thereafter, in the early 1980s, an agricultural anthropologist, Jeffrey Bentley, completed a detailed ethnographic study of peasants and landscape change in Pedralva, an agricultural village in northern Portugal.[13] Bentley's plot-level analyses of lands that reforested and lands that remained in agriculture provide a detailed account of the processes that drove reforestation in Western Europe during the twentieth century.

Ridges and plateaus, cut through by narrow river valleys, shape the landscape of northern Portugal, so most agriculture occurs on sloped lands. The Mediterranean climate delivers copious amounts of rain (2,000 mm annually) to croplands in winter when plants do not need the moisture. No rain falls during the summertime growing season, so farmers had to find ways to irrigate their crops if they were to reap life sustaining harvests from the land. Of the crops grown in Pedralva, corn, squash, beans, rye, and grapes provided the steadiest economic returns. Extreme land fragmentation characterized the agricultural sector. Sixty-eight percent of the farms in Pedralva had less than 0.5 hectares of land. Even farms this small often had several plots of land. The larger farms, between 4 and 8.5 hectares, contained on average 11.5 plots of land![14] Plots as small as 30m by 30m would get attention from cultivators, presumably because in Pedralva's impoverished agricultural context even the small harvests from these fields had a significant impact on a household's welfare. The large farms had the most fragmented landholdings. Farmers would spend large amounts of time in transit between the different plots of land on their farms.

The lengthy total transit times between plots made landowners very aware of just how much labor they had to expend in order to coax a substantial crop from one of these small, sometimes remote fields. Each plot, wherever it was, absorbed considerable amounts of labor. Farmers would terrace the sloped land to capture moisture from the rains. To irrigate a field, Pedralva smallholders would dig wells, an arduous undertaking in which farmers working with hand tools had to cut through layers of granite to reach water. Farmers would fertilize their fields with a mulch made from brush from the surrounding forest mixed with manure from animal stalls. In sum, the peasants of Pedralva practiced a labor intensive, capital scarce agriculture through the first six decades of the twentieth century that frequently, despite their best efforts, left them living in a gut-wrenching poverty.[15]

Reforestation began in Pedralva and the surrounding communities during the early twentieth century. Farmers began to abandon small plots of land located a long distance from their homes. While the abandoned plots did not have particularly accentuated terrain, they usually did not have a reliable source of water to irrigate crops.[16] When reforestation did occur in these settings, it usually took the form of spontaneous regeneration, although in some instances landowners planted eucalyptus (*Eucalyptus globulus*) in reforesting sites. This profile of reforesting plots of land fits the description of reforesting land in Alexander Mather and Carolyn Needle's explanation for reforestation through land abandonment.[17]

The drivers of reforestation in Pedralva changed in 1964 when France, along with other affluent European nations, liberalized their laws regarding the immigration of foreign workers. This change in immigration laws expanded the wage labor opportunities available to the men and women of Pedralva. They could now go overseas and earn much higher wages. Large numbers of men and women, as many as one from every other household, emigrated to France from Pedralva during the 1960s and 1970s. Their departure made wage labor more expensive in the village, and in doing so it altered the economics of working remote fields. Given the labor-intensive nature of agriculture in remote, relatively unproductive fields, it made little sense to continue working these fields when alternative economic opportunities became available. For the larger farmers, the increased difficulty of finding and contracting with farm workers after 1964 added to the reasons for abandoning the cultivation of remote fields.

When landowners decided to allow a field to revert to forest, they decided in effect to alter the array of products that they took from the land. They opted for forest products rather than agricultural products because the production of the forest products on these plots was much less labor-intensive than the cultivation of agricultural products. In a context where farm labor had become more difficult to procure, opting for less labor-intensive land uses such as forests suddenly made economic sense. When French legislators liberalized France's regulations regarding foreign workers, they increased the opportunity costs of agriculture in the migrant sending regions of the Iberian Peninsula. Under these circumstances, land abandonment accelerated in these regions. Farmers abandoned the dry, small, remote, low-return agricultural fields. The residents of Pedralva took to saying, "Today there is no misery." The reforesting fields signaled their newly prosperous condition.[18]

The earnings of the emigrants from Pedralva also began to have an impact on land uses in the village. Emigrants would send remittances to their families. Alternatively, they would return to the village after an extended stay abroad and build a retirement home, either in the village or a nearby city. The remittances from the emigrants spurred reforestation by making farm families who received the remittances more reluctant to work on local farms. Similarly, the return of emigrants to northern Portugal and their expenditures in the non-farm sector in northern Portugal created non-farm jobs that pulled smallholders off the land and encouraged the reforestation of fields that they had cultivated.

In a stratified agrarian community like Pedralva, large landowners would have the greater incentive to abandon small, remote fields and allow them to return to forest. Large landowners valued their time more than did small landowners because, with their assets, large landowners had more alternative economic opportunities than small landowners. For this reason, large landowners found the low economic returns from cultivating these fields especially undesirable and were more likely for that reason to abandon these lands. Poorer farmers with less land and fewer alternative economic opportunities found the lower economic returns from continued work in these marginal fields more tolerable.

Common trends ran across agricultural activities in Pedralva and in agricultural communities in other European societies during the post–World War II period. Farmers in other Western European societies, like those in Pedralva, worked fragmented landholdings. Over time, some consolidation in landholdings occurred in those countries, so, in both Pedralva and other Western European agricultural communities, fields got bigger.[19] Farmers purchased agricultural machinery to work these fields, applied more chemical fertilizers to their fields, and harvested timber from their woodlots more frequently. Taken together, these changes represented agricultural intensification. They increased agricultural productivity in northern Portugal and, more generally, in Western Europe. At the same time, European societies, including Portugal, began purchasing more foodstuffs from new, very productive agricultural areas overseas in Brazil and Indonesia.[20] These new, overseas sources for agricultural commodities probably precipitated additional waves of land abandonment in Western Europe and northern Portugal. These parallel trends between northern Portugal and Western Europe suggest that the drivers of reforestation in Pedralva would characterize other rural communities in Western Europe. Absent detailed comparisons between Pedralva and other rural communities in Western Europe, this conclusion about common patterns of reforestation during the twentieth century remains tentative.

Some evidence suggests that reforestation in northern Portugal occurred for a historically delimited time. One recent study of forest cover trends in northern Portugal found that the spread of wildfires after 1990, a probable consequence of climate change, destroyed forests and left burnt-over landscapes dominated by fire-resistant shrubs in their wake.[21] As in the northern Appalachians, northern Portugal experienced a historically delimited wave of reforestation that came to a sudden end. In the northeastern United States, the forests persisted even though they stopped expanding. In northern Portugal, the fires stopped the further expansion of the forests and destroyed many of the recently emerged forests.

"Cotton Fields No More": Land Abandonment and Reforestation in the American South, 1935–1975

The United States experienced a second surge in reforestation during the mid–twentieth century when large numbers of impoverished farmers abandoned agriculture in the southern region of the country. Between 1935 and 1975, according to forest inventory data collected by the United States Forest Service, approximately 9 percent of the South reforested. The abandoned agricultural lands concentrated in two regions of the South, in the Piedmont with rolling terrain in the southern interior and, just to the west, in the more accentuated terrain of the southern Appalachians. Rainfall is abundant in the South. It ranges from 1,000 mm to 1,750 mm per year throughout the South. Particularly humid regions occur along the Gulf of Mexico and in the most mountainous districts of the southern Appalachians. When the rains fall during the summer months, they often come in the form of thunderstorms. These torrential rains fall with great force, which in turn accelerates soil erosion, especially in the row-cropped cotton fields that predominated in southern agriculture during the nineteenth and twentieth centuries.

After displacing Indigenous peoples during the eighteenth and the first part of the nineteenth century, European settlers dedicated themselves to the production of cotton for global markets using slaves to do the strenuous stoop labor of cultivating the cotton. The Civil War emancipated African Americans and destroyed much of the wealth of the white landowning class. Reconstruction failed to provide African Americans with any resources. They had only their labor to sell. Under these circumstances, white landowners, newly freed African Americans, and poor whites reorganized themselves into a sharecropping system in which landowners "furnished" tenant farmers with foodstuffs for the winter and then for the growing season while the tenants cultivated a crop of cotton. For their labors, the tenant farmers received a share of the crop. Because landlords charged the tenants exorbitant prices for food, the tenants found it nearly impossible to amass the savings necessary to buy land. They were trapped. Over time, their economic conditions deteriorated. The percentage of

African American farmers who owned land declined from 25 percent in 1900 to 20 percent in 1930. The proportion of tenant farmers among all farmers in Georgia increased from 41 percent in 1900 to 60 percent in 1930.[22]

By the beginning of the twentieth century, a "cotton belt" of farms devoted exclusively to the cultivation of cotton stretched more than a thousand miles across the South, from Texas to Virginia. The repeated cultivation of row crops such as cotton in a region with frequent hard rains eroded the land and washed nutrients from the soils. The first aerial views of this landscape, taken during the 1920s, revealed expanses of red soils with only a thin vegetative cover. The torrential rains on these bare soils had created networks of red, chasm-like gullies, dozens of feet deep, that stretched across the cotton belt. Gullies scar the land in figures 4.3 and 4.4. They made much of the land unworkable.

Farmers tried to combat soil exhaustion through liberal applications of chemical fertilizers to cotton fields. In 1929 farmers in the Southeast spent $2.71 on fertilizers per acre compared with expenditures of $0.30 for

FIGURE 4.3 Eroded fields, Mississippi, 1936. Photo courtesy of U.S. Department of Agriculture, 1936.

FIGURE 4.4 Abandoned and eroded agricultural land in a national forest. Photo courtesy of U.S. Department of Agriculture, 1936.

fertilizers per acre by farmers in the Midwest. The cost of fertilizers equaled one-third to one quarter of the annual value of the cotton crop in South Carolina by the late 1920s.[23] These expenditures placed southern cotton farmers at a competitive disadvantage compared with cotton cultivators in other, expanding production zones such as California and Egypt. Invasive species, in particular the boll weevil (*Anthonomus grandis*), added to the farmers' troubles. The boll weevil spread northeastward during the 1920s, wiping out cotton crops in many counties and adding to the economic burdens faced by cotton sharecroppers from Texas to Georgia. In the early 1920s, the arrival of the boll weevil in a Georgia county precipitated a 70 percent decline in the size of the cotton crop.[24]

Despite the accumulating economic burdens shouldered by cotton sharecroppers after 1900, they continued to produce substantial amounts of cotton, in large part because its price stayed high through 1920, buoyed by the large wartime demands for cotton from the combatants in World War I.

When the demand for cotton dropped at the end of the war, the price of cotton collapsed, and it remained very low throughout the 1920s. The low price of cotton, the degraded state of agricultural lands in the cotton belt, the continuing economic exploitation of tenant farmers, and the persecution of African Americans by the Jim Crow system spurred a large-scale out-migration from the South by African American and white tenant farmers during the 1920s.[25] The landowners, without a labor force, could not produce a crop, so they did not cultivate their lands and defaulted on their loans. By 1930, banks and insurance companies had become the owners of as much as 30 percent of the land in some Black Belt counties in Georgia.[26]

At the same time that sharecroppers were abandoning the land, the mechanization of agriculture freed up additional lands for reforestation. Tractors replaced mules and horses on farms. With little need for mules and horses, pastures no longer seemed necessary. The low prices for cotton, the unworkable, gully-damaged fields, the surplus of pastures, and the outmigration of tenant farmers all contributed to an increase in abandoned agricultural lands across the southern cotton belt during the 1920s. Given the abundant rainfall in the region, trees, most often pines, sprouted in the old fields.

Through this dynamic, reforestation in the South had already commenced when the Depression began in 1929. The economic impact of the Depression on Americans cannot be overstated. From the beginning of the Depression in 1929 to its bottom in 1933, the gross national product of the United States declined by 29 percent. Consumption expenditures declined by 18 percent, construction by 75 percent, and investment by 98 percent. The unemployment rate soared from 3.2 percent in 1929 to 24.9 percent in 1933.[27] The losses of jobs in cities spurred a return migration to farms among white Southerners. Between 1930 and 1935, the population on farms increased by 1.3 million persons. Agriculture could only support the return migrants at subsistence levels. The price of cotton remained depressed. In the words of Henry Wallace, the secretary of agriculture during the 1930s, destitute peoples were "piling up" on farms.[28]

When Franklin Delano Roosevelt became president of the United States in 1932, he and his advisors formulated a series of policies to address the Depression-accentuated problems of rural poverty. These policy changes had important impacts on land abandonment and reforestation in the cotton belt. The most significant of these policies, called the Domestic Allotment program, sought to raise the prices of agricultural commodities by limiting their production. Cotton farmers would, for example, receive a subsidy from the government if they promised to reduce the extent of their cotton fields by 15 percent. The reduced acreage devoted to cotton would diminish the size of the harvest, thereby raising its price. The higher prices for cotton would in turn reduce the poverty of landowners.

Farmers could choose which of their lands to retire. Not surprisingly, they chose to retire their least productive lands, and these lands over time became secondary forests. The productivity of the lands that remained in cultivation would of course rise, if only because they constituted a more selected set of croplands. The land use changes in twelve cotton-growing Mississippi counties typified the larger regional pattern. Growers in the twelve counties reduced their cotton acreage from 700,000 hectares in 1930 to 400,000 hectares in 1940. Over the same period, cotton yields grew from 220 to 440 kilograms per hectare. In 1933, the first year of the allotment program, administered by the Agricultural Adjustment Administration (AAA), cotton growers retired more than 4,200,000 hectares that had been cultivated the previous year.[29]

Other, smaller programs contributed to the same goal of returning marginal lands to forest. The federal government initiated a buyback and resettlement scheme that purchased degraded lands from small farmers. The lands purchased from bankrupted growers often became parts of a network of National Forests in the South established by Roosevelt. Figure 4.4 captures this seemingly paradoxical outcome: degraded croplands within a newly established forest preserve. In several instances, the bought-back lands became bases for resettlement schemes in southern Appalachian valleys. Related programs subsidized the purchase of tree seedlings by farmers and provided laborers from the Civilian Conservation Corps to plant the seedlings on abandoned farmlands. Both the buyback and the reforestation subsidy schemes affected relatively small areas of land compared to the allotment program.

While the New Deal programs accelerated the abandonment of cotton fields, the reversions of these fields to forests continued for decades after the 1930s, fueled in part by the continuing increase in the productivity of the crops being grown in the region. Largely through more liberal applications of fertilizer, crops such as cotton, tobacco, corn, and peanuts all saw increases in crop yields of more than 100 percent between 1935 and 1965.[30] The magnitude of these increases in the land productivity of crops such as cotton enabled reductions in the area planted. Through this dynamic, cotton cultivation retreated over a thirty-year period from the expansive cotton belt extending from Texas to Virginia to a single block of reclaimed, alluvial, and very fertile soils in the Mississippi Delta. These reclaimed lands became available for cultivation through the construction of federally funded levees in the Mississippi Delta after the devastating flooding of the Mississippi River in 1927.[31] The abandoned cotton fields elsewhere in the South became either pastures or forests.

Did the new forests persist into the twenty-first century? Only in part. The resurgence of secondary forests in accentuated terrain, in southeastern Oklahoma for example, has continued in a few places, but large portions of the forests in the former cotton belt have disappeared as metropolitan areas in the Piedmont expanded during the second half of the twentieth century. Suburbs outside of Atlanta and Charlotte grew rapidly after 1970. The prevalent pattern of urbanization involved the construction of subdivisions of single-family homes that occupied large areas of land and require the clearing of extensive amounts of secondary forest.[32] In this respect, the gains in forest cover, at least in the Piedmont region proved to be temporary, like the forest gains in northern Portugal and the earlier forest gains in the northern Appalachians. Nonetheless, for at least three decades, from 1935 to 1965, the American agricultural sector, with multiple organizations representing mostly prosperous farmers and policymakers, spurred, through a corporatist process, the reforestation of a significant portion of the American South.

The Decline of Cattle Ranching and Forest Resurgence in Costa Rica

Between 1975 and 2000, the Tempisque River Basin in the Guanacaste region of northwest Costa Rica experienced a dramatic increase in secondary forests. Grasslands for pasturing cattle declined from 290,000 hectares to 150,000 hectares at the same time that forests increased from 155,000 to 256,000 hectares.[33] This surge in reforestation occurred in a dry tropical landscape. Rainfall is abundant but seasonal in its distribution, with almost all of the precipitation falling between May and November. The terrain features a mountain range, the Cordillera de Guanacaste, that rises to 2,000 meters above sea level in the east and then slopes down to the Tempisque River and the Pacific coast in the west.

Colonists claimed large chunks of this landscape, deforested it, and planted pasture grasses for cattle during the nineteenth century. The owners and managers of these haciendas established a transhumant grazing routine. During the dry season, they would take their semi-feral herds of cattle up into the mountains where showers maintained adequate forage. The introduction of African pasture grasses after 1880 and Brahman cattle after 1920 increased the landowners' investment in their herds. To prevent the loss of these resources, hacienda owners began to fence off the pastures that they had sown and the new cattle that they had purchased.

These enclosures stripped subsistence-oriented peasants of access to lands for gardens and for pastures for their livestock. The concentration of pastures and cattle in a few enterprises encouraged the landless peasants in the region to migrate elsewhere, usually to cities, during the 1960s and 1970s.[34] Despite the new forages and new breeds of cattle on some ranches, extensive cattle ranching remained the defining feature of agricultural livelihoods in northwestern Costa Rica during the 1960s and 1970s. The pastures supported less than one head of cattle per hectare. Ranchers, long resident in the region and organized into an association, wielded considerable political power in the national capital. Through the exercise of influence, they secured public subsidies such as generous lines of bank credit that at times provided loans with negative rates of interest.[35] The cattle ranchers exported substantial quantities of beef to the United States despite the extensive, decapitalized condition of their haciendas. By the late 1970s, Costa Rica ranked fourth by weight among the exporters of beef to the United States.[36]

Beginning in the 1970s, price trends for cattle began to hurt the Guanacaste ranchers. Beef prices that had averaged $2.37 per kilogram during the

1970s declined to $1.36 per kilogram during the 1985–1999 period. Declines in per capita consumption of beef in the United States, stemming from increased consumer concern about beef consumption and heart disease, most likely explain the price declines.[37] The international debt crisis of the early 1980s worked in tandem with the price declines in beef to throw cattle ranching enterprises into a crisis. A structural adjustment agreement between the International Monetary Fund (IMF) and the government of Costa Rica eliminated the subsidies that the national government had provided to cattle ranchers. At the same time, the government financed the construction of 234 kilometers of irrigation canals in the Tempisque River basin. The canals made it possible to irrigate crops in the lowlands of the basin during the dry season. In so doing, the network of new canals, coupled with the low prices for cattle and the loss of government subsidies, encouraged landowners to transition away from cattle ranching in the Guanacaste region.[38] They abandoned their upland pastures which then reverted to forest.

While ranchers abandoned lands in the Guanacaste mountains, they intensified their use of land at the lower elevations where they dug irrigation ditches from the Tempisque River. The construction of the canals enabled irrigated agriculture where it had not existed before. Intensification also occurred on those lands that remained in pasture. The numbers of cattle went up in the Tempisque Basin during the same period that the amount of pasture in the district declined by almost 50 percent.[39] Agricultural intensification accompanied reforestation and, arguably, spurred it. Protected areas expanded in extent throughout Costa Rica during the 1980s and 1990s as part of an effort to attract more ecotourists to Costa Rica.

The expansion of protected areas in the Guanacaste region, in particular in the mountainous regions, included some pastures. These lands, now protected, reverted to forests during the first decade of the twenty-first century. The creation of one of the world's first payment for environmental services (PES) programs in Costa Rica in 1996 only played a small role in the reforestation, largely because few landowners in the Guanacaste region participated in the program's initial phases.[40]

Another more diffuse but pervasive influence on land abandonment and reforestation stemmed from the shift of the foundations of the Costa

Rican economy from agrarian livelihoods to service-based tourism. After 1994, tourism became the largest economic sector in Costa Rica. This shift had palpable effects on individual households. Younger people moved off of the farms and into towns and cities where, most often, they took jobs in the non-farm service sector. The rural–urban migration created labor scarcities on farms that induced farmland abandonment in some locales. As one old farmer in a district just to the south of Guanacaste put it, "We have people to inherit the farms; we just do not have anyone to work them!" Remote sensing analyses of land use changes in this district show a similar pattern to the one in Guanacaste. Reforested lands tended to occur on slopes and near watercourses. Farmers saw emerging secondary forests along streams as a way of protecting sources of water for their livestock.[41]

The reforestation in the dry forests and the mountains of northwestern Costa Rica occurred in those parts of Costa Rica least suited for agriculture. In other words, agriculture in Costa Rica persisted in those parts of country most suited to it. Viewed transnationally, the land use changes in Costa Rica, led by the changes in Guanacaste, had globally beneficial environmental effects. The land area that went back into forest in Costa Rica amounted to 130 percent of the area in developed countries such as the United States that replaced the lost production from the reforested Costa Rican places. In this sense, the Costa Rican reforestation contributed to the concentration of agricultural production globally in those places most capable of producing food efficiently.[42]

The environmental benefits of restoring forests depend on the long-term persistence of the restored forests. Research on reemerging forests in southern Costa Rica raised questions about the persistence of the expanded forests. The setting involved upland regions above 700 meters with annual precipitation of 3,500 to 4,000 mm. From 1950 to1980, smallholders carved up these forests to create coffee plantations and then allowed some of the older coffee groves to revert to forests. Remote sensing analyses indicate that landholders cleared 50 percent of the recovering forest patches within the first twenty years after the forests reemerged.[43] While these rates of new forest destruction would make natural climate solutions ineffective, it is important to consider the ecological settings in which the reforestation occurred. The reforestation in northern Costa Rica occurred in dry forested areas with few alternative economic activities. In this respect, the impetus to reclear newly emerged forests may have been stronger in the more humid, agriculturally productive southern zones than it was in the more arid, rugged terrain of northwestern Costa Rica. Certainly, the strong advocacy of naturally regenerated secondary forests, as the preferred means for forest resurgence, presumes that people will allow the new forests to persist and, over the decades, become old growth forests.

General Patterns in the Spread of Secondary Forests

Several persistent trends in contemporary societies appear to encourage the spread of secondary forests across landscapes. Table 4.1 catalogs these patterns. Urbanization, agricultural intensification, degraded mountain landscapes, and large landholdings associate persistently with the spread of secondary forests.

Urbanization, Agricultural Intensification, and Land Abandonment

Of the hypothesized contributors to land abandonment in table 4.1, only one driver, rural to urban migration, appears in all four cases. Clearly, urban economic and demographic growth encouraged the abandonment and

TABLE 4.1 The Contexts for Land Abandonment and Secondary Forest Expansion

	Mountain terrain	Soil exhaustion	Falling commodity prices	Agricultural intensification elsewhere	State initiatives	Rural–urban migration
New England	Yes	Yes	No	Yes	No	Yes
Portugal	Yes	No	No	No	No	Yes
Southern United States	No	Yes	Yes	Yes	Yes	Yes
Costa Rica	Yes	No	Yes	Yes	No	Yes

Sources: Case studies in chapters; see notes.

reforestation of old fields in rural places. The growth in urban industrial enterprises and the accompanying expansion in the size of urban populations triggered changes in land uses extending outward from the cities. The growing human population put pressure on the food supply, so farmers extended the area of land that they cultivated. For similar reasons, loggers extended their activities into more remote, as yet uncut forests farther from center cities. Over time, farmers got to know their now more extensive croplands better. Each harvest brought farmers new knowledge about the productivity of their lands, and farmers began to draw conclusions about the relative merits of the different fields on their farms.[44] Farmers learned from their observations to focus their efforts on their most fertile lands. In doing so they managed to maintain a constant volume of production while working a smaller area of land.[45]

While farmers expanded cultivated areas to meet the rising demands for foodstuffs from growing urban populations, farm workers began to consider new employment opportunities in urban areas, in particular in the newly established factories in cities. In all four of the case studies, farm workers, attracted by the higher wages available in factories, decided to leave the farms and move to cities. Farmers faced a dilemma when they lost workers to cities. To retain enough workers to cultivate all of their fields, farmers had to raise the wages of their workers. With wage labor now costing more, some farmers found it unprofitable to cultivate the least productive of their fields, so they abandoned agriculture on these fields. Quite frequently, the abandoned fields occupied sloped land in relatively inaccessible locations. As noted in the Portuguese case, remittances from the emigrants affected the pace with which secondary forests reoccupied abandoned farmlands. The outside income from the emigrants reduced the willingness of the remaining laborers in a community to work on landowners' fields.

In these settings, outmigrants encourage land abandonment both by leaving the land and by remitting some of their earnings to family members who continue to reside in the sending regions. This pattern seems quite widespread. In addition to the cases described in this chapter, it has been documented in Nepal, Central America, and the Caribbean.[46] A variant on this larger pattern occurs when the remittance-sending, labor migrants return from overseas. They frequently engage in a kind of "boomerang migration." When the migrants left the region, they left small, out-of-the way farms in search of work, and the fields on their farms returned to forest. When they returned to the region from overseas twenty or more years later, they did not return to the farm that they had left. Instead, they usually went to the nearest large town where they used their accumulated savings to buy real estate or start a business.[47] The labor migrants left, and, like a boomerang, they returned, but not to quite the same place. Given this pattern, the return of the labor migrants did not reverse the succession of old fields into secondary forests in the migrants' sending region. By focusing the new investments on urban but not rural areas of sending regions, the labor migrants' return may have induced an additional wave of departures from the outlying farms and spurred additional abandonment of peripheral croplands. The geographical pattern of this abandonment dynamic follows economist Johann Heinrich von Thünen's central place theory, with land abandonment and the associated reemergence of forests occurring more frequently in relatively remote, rural settings where the inaccessibility of the land makes alternative economic uses of the land difficult, if not impossible.[48]

Forests beget more forests in this pattern of change. Seed rain from nearby patches of forest enables tree seedlings to establish themselves on the abandoned croplands. The seedlings initiate the processes of old-field succession that reestablish secondary forests on croplands. The economic logic that underlies this pattern of old-field succession conveys an image of reforestation in increments. It occurs as a press process, through slow, incremental gains spread out over time.[49] Price trends for labor, land, and agricultural commodities make it impossible to cultivate fields profitably, so farmers decide in an incremental way, one field at a time, to abandon land.

In three of the four cases described earlier—New England, Portugal, and the southern United States—farmers cultivated two staple crops, corn and cotton. When farmers increased production and prices declined, consumers did not respond by buying more of the staples, so their prices declined, and, in response, farmers abandoned their most unproductive fields.[50] Land sparing occurred. A more difficult to discern pattern of land sparing distinguishes between local and extralocal effects of intensified agriculture on decisions to abandon cropland. As farmers observe their neighbors' success at intensified cultivation and decide to do likewise, intensified production and additional deforestation would occur locally. A contrary effect would occur extralocally. The extra production from the recently intensified farms, if large in magnitude, would depress the global price of the relevant agricultural commodities. Farmers in districts far from the intensified farms would experience a disincentive to farm from the lower commodity prices without experiencing the opposed incentive to expand farm operations that comes from observing the success of their neighbors at intensified agricultural operations.[51]

As suggested by this dynamic, the globalization of markets for agricultural commodities during the second half of the twentieth century had broad regional effects on the drivers of secondary forest expansion. Wealthy countries such as France imported a larger proportion of their foodstuffs from inexpensive, intensifying, overseas producers in humid, agriculturally productive, tropical countries such as Brazil and Indonesia. The increased reliance on these overseas sources of food made it difficult for local European farmers to turn a profit on the production of crops and livestock from their least productive fields. Disparities in the price of labor between wealthier countries in Europe and poorer countries in the Global South added to the economic incentives to abandon agriculture in the wealthier countries. Under these circumstances, French farmers would allow some of their unproductive, usually upland, fields and pastures to revert to forest.[52]

Larger Farms and Land Abandonment

The overall size of landholdings mattered in decisions to abandon land. Smallholders often worked all of their land in an effort to meet the subsistence needs of their households. With more land than they needed to meet their subsistence needs, large landowners tended to think in terms of the marginal economic returns from fields rather than in terms of the impact of a field's harvest on the household's subsistence. On the bases of these comparisons, large landowners allowed some fields but not others to lie fallow for one or more years. The fallowed fields then became potential sites for regenerating forests if the income streams from the still-cultivated fields satisfied farmers.

A landowner's sense of security about his or her title to land also shaped decisions about forest clearing and regrowth. Insecure land tenure has characterized most small and medium sized tracts of forests in the Global South. Farmers would clear forested land out of the belief that the visual evidence of having "worked" the land, as in figure 3.1, would strengthen their claim to it.[53] In this manner, insecure land tenure accelerates deforestation in frontier zones. It also retards reforestation. Smallholders see cleared land as evidence for their presence on the land. A neighbor who covets a cleared tract of land knows that someone has occupied it and that an invader will encounter resistance if she or he were to occupy the property.

Large landowners have been more likely to have secure titles to land. Beginning in the 1960s, states established programs to title land, and, after 1980, an increasing number of these programs began to charge fees for titling the land that many smallholders could not pay. Under these circumstances, large landowners typically had more secure titles to their land than smallholders because large landowners could pay for a full title to the land. The more secure titles to these larger tracts of land in turn encouraged larger landowners to leave their land in forest if they felt that the anticipated yields from particular tracts of land did not justify land clearing.[54]

Evidence for this pattern of more forest retention or reforestation on larger ranches and farms comes from diverse sources. As noted in chapter 2, large landowners on several Amazonian frontiers in Brazil retained much more of their land in forest than did small landowners.[55] The politically connected accumulation of large landholdings in sub-Saharan Africa (SSA) appears associated with a willingness to let the land lie fallow. In one survey in Zambia, 90 percent of the land in the 20–100 hectare category, a large farm category in SSA, remained uncultivated and likely to regenerate into woodlands.[56]

The growing size of the remaining farms enables the landowners to pursue economies of scale that in turn increase the volume of production coming from the remaining fields and depress the prices of staple crops, such as wheat or corn, produced on these farms. The lower prices of agricultural commodities in turn provide an additional impetus for smallholders to abandon agriculture on small, topographically disadvantaged farms. Through this indirect mechanism, the large, intensified farms save old growth forests from destruction elsewhere and induce the spread of secondary forests onto relatively unproductive croplands.[57] In other words, intensified agriculture spares less productive agricultural lands from the degrading effects of cultivation.[58] Payments for environmental services in this context could end up delivering the largest payments to the largest landowners, so compensatory payments to conservation-minded smallholders would seem justifiable as a way to address the inequities associated with payments for carbon offsets.

Mountainous Districts and Land Abandonment

While the natural regeneration in New England, northern Portugal, and Costa Rica all occurred in montane settings, the reforestation in the southern United States occurred in a rolling landscape that, at least initially, would have been workable by plow and horse and later would have been workable with tractors. Perhaps in this respect the cotton belt is the exception that proves the rule that only montane landscapes undergo frequent reforestation. Large parts of the once-arable lands of the Piedmont region, had, through erosion, evolved by early in the twentieth century into such an extensive network of deep gullies across the fields as to make them impossible to cultivate. Like many small fields in montane locations, the old, eroded fields in the cotton belt had become impossible to plough with either horses or machinery. Figures 4.3 and 4.4 depict the unworkable terrain. Under these circumstances, land abandonment by cotton farmers seemed like a reasonable course of action and a spontaneous resurgence of secondary forests occurred. In the following chapter, figure 5.1 makes the same point with a quite different example. The grid that organizes mechanized oil palm cultivation follows the contours of the land almost exactly. The strong association between techniques of cultivation and topography indicates how much the shape of the terrain in a place influences the likelihood that particular tracts of agricultural land will revert to forest.

Topography interacts with regional agricultural economies in ways that produce variations in secondary forest regrowth. Smallholders, at least in Latin America, have clustered in mountainous topography.[59] These places become sending regions for rural–urban migrants. The departure of the migrants precipitates the abandonment of the sloped fields that predominate in these regions. In lowlands with more even terrain and better road networks, large-scale farms predominate. When forest restoration does occur in lowland settings, it usually involves one or more fields with unusual drainage issues.

The differences in slope between abandoned and cultivated fields are often not extreme. As a result, intensely cultivated regions with only modest differences between farms in topography can have large numbers of fields on one farm that will lie fallow and potentially begin to reforest in a given year while on a neighboring farm very few fields lie fallow. For example, substantial amounts of reforestation occurred in southern Illinois, an intensified agricultural region where gentle, rolling hills reach a little more than 300 meters above sea level. Despite the relative uniformity of the upland, transitional, and bottomland terrain in this region of Illinois, the upland portions saw an increase in forest cover from 10.9 percent of all land in 1938 to 19.1 percent in 1993. During the same period, bottomland forests decreased from 19.0 percent in 1938 to 11.4 percent of all land in 1993.[60] The reemerging forests have concentrated in the hilly terrain.

The concentration of land abandonment in the highlands has occurred across a wide range of places, including three of the four case studies in this chapter. It has also been documented in Myanmar, in Central America, in Europe, and in Argentina.[61] A highlands focused pattern of forest expansion has implications for the preservation of biodiversity because sloped land typically contains more niches for species than flat lands.[62] Given that sloped land usually contains more biodiverse flora, a pattern of upland farm abandonment would have positive implications for the preservation of biodiversity.

The Role of the State in Land Abandonment

States have, on occasion, played important roles in spurring the natural regeneration of forests. As noted earlier in this chapter, government-supported restoration efforts began with the New Deal in the United States. The Agricultural Adjustment Act of 1933, reinforced by the Conservation Reserve Program (CRP) in 1986, provided subsidies to farmers who took 15 percent or more of their lands out of production. During the late 1930s and 1940s, American farmers organized themselves into Soil Conservation Districts that reduced the acreage in crops and mandated soil conservation practices in their watersheds. In the southeastern United States, a large portion of the lands taken out of production during the New Deal reverted to forests. In the CRP, lands taken out of production did not necessarily stay out of production. A rise in the price of agricultural commodities could exceed the amount of the government subsidy for that year, in which case U.S. farmers could raise their income by converting their conserved lands back into croplands for that year. The short-term gains in the extent of forests through these fluctuations would not produce the long-term gains in carbon sequestration necessary to counter climate change.

In 2013 the European Union initiated a government-supported restoration effort that promises to achieve more long-term gains in carbon sequestration. It required that farmers establish 5 percent of their land area as an Ecological Focus Area (EFA) and take it out of production. In return, the farmers would receive subsidies through the European Union's Common Agricultural Policy (CAP). EFAs would typically take the form of hedges or woodlots that would increase the biodiversity on a farm or in an agricultural sector. Unlike the CRP regulations in the United States, the 2013 reforms of the CAP did not build in an opt-out provision for farmers in the event that crop prices rose during a particular year.[63] Without this provision, the EU policy provides a more secure way of mitigating greenhouse gas emissions than the CRP in the United States. While both the arrangements in the European Union and the United States would seem to qualify for a REDD+ system of payments for carbon offsets to cooperating farmers, neither farmers nor climate activists have pursued this type of reform.

Both of the reformed sectors in the European Union and the United States exhibit the attributes of societal corporatism. They both maintain high levels of organizational density. Most farmers in the European Union belong to cooperatives that assist them with the purchase of inputs and the sale of harvests. The associations of cooperatives within nations integrate across national boundaries, so farmers have influential voices in the construction of EU-wide agricultural policy. Large-scale American farmers, through organizations such as the Farm Bureau, stake out policy positions and wield considerable amounts of power in national agricultural policymaking. With their highly organized bases, their effective representation at the highest levels of policymaking, and their sustained interactions with environmental policymakers, both the American and the European agriculture sectors exhibit the attributes of societal corporatism. While a sequence of political events that would lead to state-supported land abandonment becomes conceivable in this context, the absence of a substantial commitment to the climate stabilization movement among American farmers makes forest gains an unlikely outcome of the sector's corporatist processes.

In some instances, disasters have provided the impetus for government reforms. Floods in nineteenth-century France and Switzerland provoked a response from the state that jumpstarted reforestation in the mountains. Dramatic, downstream floods after heavy rains in deforested highlands underlined the costs to downstream regions of maintaining croplands in the surrounding highlands. In the aftermath of these disasters, governments created incentives for highland farmers to reforest some, if not all, of their lands, and these changes in policy expedited reforestation.[64] Politicians created a corporatist structure that linked farmers, foresters, and downstream urban politicians in a collective effort to alter highland landscapes in order to prevent damaging floods in downstream cities.

In effect, dramatic events reordered the legislative priorities of rulers, and they passed legislation that changed the incentives surrounding agricultural land abandonment. The legal framework or "regime" surrounding upland agriculture shifted, and a surge in land abandonment occurred.[65] Viewed historically, this type of old-field reforestation has occurred unevenly, with a period of slow conversion to forests followed by a surge in the rate of reforestation after a focusing event altered the political and economic calculus of the farmers' operations.

While collective action following a natural disaster explains a number of well-known shifts from deforestation to reforestation, political convulsions, unrelated to forests, have sometimes produced substantial land abandonment and forest resurgence in the short-term. The collapse of the Soviet Union in 1991 and the subsequent reforestation of prime agricultural lands offers a case in point. In this instance, the government disintegrated amid a political crisis, and government subsidies for collective farms declined dramatically. Under these circumstances, workers on the collective farms reduced the amounts of land that they cultivated, and the abandoned croplands reverted to forests and scrub growth. Unlike settings where farmers abandoned marginal croplands, these farmers let prime farmlands revert to forest.[66] After the political system stabilized in the 2000s, farmers resumed cultivating some of the prime farmlands.[67]

A similar sequence of events occurred in Cuba in the 1990s when the disintegration of the Soviet bloc of nations deprived the government of its preferred overseas market for sugar. In this instance, sugar mills shut down and an invasive shrub, marabou (*Dichrostachys cinerea*), spread across the now unused plantation fields.[68] This shift in agricultural policy altered landscapes in Cuba. The state reduced its expenditures in the sugar sector, and sugar plantations declined in size in remote rural districts. At the same time, the state began to promote rice and bean cultivation by periurban smallholders near Havana, so cultivated areas around cities remained unchanged.[69]

Can We Expect a Continuing Spread of Secondary Forests?

Three of the four case studies of forest restoration in this chapter concerned places that experienced large-scale agricultural expansions in the nineteenth and twentieth centuries. The later forest resurgence in these places almost has a tidelike quality to it. The human tide came in when poor farmers occupied all of the arable land in a region during the eighteenth and nineteenth centuries. Then it went out. After a succession of growing seasons, a winnowing or selection process began. Poorer farmers on infertile lands gave up agriculture, and these lands returned to forest. This migration away from the land has now run its course in places such as North America, so it seems reasonable to ask how many other places on the globe might experience substantial rural outmigration, agricultural intensification, land abandonment, and forest resurgence. Would the coronavirus pandemic with its high levels of contagion reduce rates of rural to urban migration and, in so doing, diminish the extent of abandoned lands and forest recovery?

A second issue concerns the durability of secondary forests. Forests as a natural climate solution presumes that, when secondary forests reoccupy a landscape and absorb carbon from the atmosphere, the trees will be left standing for a half-century or more, sequestering carbon. Several studies in Latin America have found fault with this assumption. They report that, controlling for other conditions, secondary forests are chopped down more frequently than are old growth forests.[70] This pattern, coupled with the declines in rural–urban migration, suggests that robust streams of sequestered carbon from expanding secondary forests will only occur with supportive interventions by governments and nongovernmental organizations (NGOs).

Finally, a cautionary note is in order about secondary forest expansion as the chief vehicle for a natural climate solution. In the old settlement zones of southeastern Brazil, the amount of the sequestered carbon appears to have declined substantially in drought-stricken secondary forests.[71] Under these circumstances, plans for large-scale reforestation should presumably rely on a range of different vehicles for reforestation. The proposals to restore forest cover by participants in the Bonn Challenge shed light on this possibility. Natural regeneration of secondary forests accounts for 34 percent of the area pledged for restoration. Forest plantations account for 45 percent of the pledged area.[72] If the limited uptake of carbon in the secondary forests of degraded, drought-stricken areas proves widespread, then climate activists would want to investigate more closely the ecological and economic potential of large-scale forest planting. The next chapter explores these possibilities.

5

• • • •

Planted Forests

CONCESSIONS, PLANTATIONS, AND THE STRENGTH OF STATES

The recent increase in tree farms, referred to as forest plantations by some observers, raises questions about associated environmental services, in particular carbon sequestration, that might come with that increase. During the initial, vigorous growth phases of recently created plantations, they approach but do not equal in their annual uptake of carbon the naturally regenerating secondary forests in the tropics. Tree farms sequester 5 t C ha^{-1} yr^{-1}, while secondary forests in the tropics sequester anywhere from 7 to 18 t C ha^{-1} yr^{-1}.[1] In a fuller accounting of carbon emissions, naturally regenerated forests usually retain much more carbon in the landscape than do forest plantations. Naturally regenerated forests avoid the additional episode of land clearing that occurs when plantation owners, in preparing the ground for planting, cut down degraded old growth forests on a site. There may, however, be a highly defined set of circumstances, beginning with already degraded sites, dominated by invasive shrubs and grasses, where tree planting would contribute to the accumulation of carbon in landscapes. Comparisons of the following case studies of plantation expansion should clarify these circumstances in which forest plantations constitute natural climate solutions. The analysis begins with a summary of the historical circumstances that have accompanied the growth of tree plantations around the world.

The History of Forest Concessions, Plantations, and the Great Acceleration

The rapid industrialization, population growth, and urbanization of the nineteenth century increased industrial demands for wood. Railroad companies wanted wood for ties on railroad lines. Coal companies wanted wood for the crossbeams that secured the ceilings in their mines. Builders wanted wood to construct two- and three-story houses for workers within walking distance of the new factories. The demand for wood to build houses increased further during the second half of the twentieth century, when builders made single family homes bigger and more expensive.

Initially, the wood for these expanding urban and industrial uses came from old growth forests. When builders and industrialists exhausted nearby sources of wood in old growth forests, suppliers turned to more distant sources. By the beginning of the twentieth century, the British coal industry, for example, had begun to meet its needs for wood by importing timber from old growth forests in Russia.[2] During the post–World War II era, the demand for wood in a reindustrializing Japan spurred the destruction of old growth forests in the outer islands of Indonesia.[3] LiDAR remote sensing techniques estimated that workers had selectively logged approximately 20 percent of all of the old-growth forests in the humid tropics in just a five year time period, between 2000 and 2005.[4] Most of this wood entered long-distance trading networks that supplied construction materials and fuel to growing urban economies in the Global North and South.

To profit from the extraction of wood from old growth forests, groups of investors throughout the tropics sought concessions that licensed them to log tracts of forested land. Later, after a first wave of cutting, many of these investors secured their own sources of supply by planting fast growing, exotic species of trees on cutover tracts of land in the concessions. By 2010, logging companies had established large forest concessions across six nations in the Congolese forest in central Africa. Other groups of investors established forest concessions across the outer islands of the Indonesian archipelago in Southeast Asia. In an initial pass through the old growth forests in these concessions, work crews practiced selective logging, extracting only two or three species of trees from forests containing a very large number of tree species. Selective logging occurred over very large areas.

As the rapid extraction of wood from old growth forests exhausted them as sources of supply, traders began to look for alternative sources of supply closer to home. As early as 1919, the British government decided to plant trees in upland areas of the British Isles as a substitute for foreign sources of wood.[5] Through these substitutions in sources of supply, forest plantations increased gradually in extent. In recent years, increases in the extent of tree farms appear to have accelerated. While tree farms only accounted for 3.2 percent of tropical tree cover in 2015, planted tree cover in the tropics did expand by 87 percent in extent between 1990 and 2015.[6] As of 2020, forest plantations constituted 7 percent of the world's forests (FAO 2020).[7]

Cities create markets for the sale of wood, so, given the considerable transportation costs of getting large volumes of timber to market, wood producers, including plantation owners, have sought accessible sites for planting trees. Beginning as early as the thirteenth century, merchants in western European cities tried to secure and preserve local supplies of fuelwood by coppicing trees close to cities. Loggers began to manage forests along rivers whose moderate flows made it possible to float logs downstream to markets in cities.[8]

Five centuries later, planters continued to prioritize accessibility in deciding where to locate tree farms. China promoted a major expansion in small forest plantations in areas with numerous road and rail links to the rapidly expanding industrial and urban core of the Chinese economy. This pattern appears in the Brazilian Amazon, where smallholders plant and harvest acai, a fruit-bearing tree, close to the city of Belem.[9] They market the fruit to Belem by boat. The same pattern of close proximity to markets characterizes tree plantations that supply wood to the housing markets in urban agglomerations such as Atlanta, Georgia.[10] Recent road building in long-deforested periurban and rural northern India led to an increase, not a decrease in forests after the road building. The new roads increased proximity to cities, with their markets for wood, which in turn spurred landowners to create forest plantations along the new roads.[11] Similarly, landowners sought coastal locations for tree plantations in Peninsular Malaysia and the outer islands of Indonesia.[12] Proximity to deepwater ports made it easier for the plantation owners to transport and sell their wood in overseas markets. Southern Brazil, near the country's industrial complexes, also saw major increases in tree plantations beginning in the 1980s.[13] All of these dynamics have clustered forest plantations in accessible locations.

Other studies have focused on the tempo of clearing on forest plantations. One recent study found that landowners cleared planted forests more frequently than they cleared naturally regenerated secondary forests.[14] The accelerated rates of clearing in planted forests, coupled with the importance of ease of access to these forests, underscores the salience of commercial considerations in land use decisions about planted forests. Plantation owners have chosen to harvest trees as often as is economically feasible.

As the close ties to urban places and the high frequency of harvesting make clear, the growing number of forest concessions and the expansion in tree plantations since the 1990s, like the expansion in tropical agriculture during the same period, have stemmed from the Great Acceleration of the human enterprise. Concession holders and plantation owners participate in a treadmill of production in which they harvest trees as frequently as economically possible.[15] These practices earn plantation owners considerable profits and diminish the environmental services provided by the growing trees.

Landowners often establish forest plantations on degraded lands consisting of logged over forests or invasive infested grasslands. Landholders spend to plant exotic trees in these places because other, easy-to-exploit economic opportunities do not exist. Only by a large investment in a plantation of exotic species can a landowner expect to reap, over time, large profits from these lands. The prospect of large, private sector profits from tree plantations may explain why so many of the Bonn forest restoration plans feature plantations as the means to a natural climate solution. Given these conjectures, it becomes useful to explore cases in which landowners, communities, and nations have chosen to expand tree farms. These analyses promise to identify the socioecological conditions in which the creation of tree farms adds sequestered carbon and other environmental services to landscapes.

The Congo: The Failure of Ecocertified
Concession Logging

Since the 1990s, groups of investors, in many instances from overseas, have applied for and obtained spatially delimited concessions in the old growth tropical forests of the Congo River basin. These concessions now cover immense areas of rain forest in Central Africa. To extract trees from these forests, concession holders construct networks of logging roads that usually end at rivers. The loggers at these sites tie together the felled tree trunks and float these rafts downstream to a sawmill. The new roads fragment the forests in the concessions. In some cases, the logging roads become overgrown and impassible within about ten years of their initial construction. In other cases, the new roads spur further changes in landscapes. Soon after the loggers build the roads and extract wood from a tract of land, families begin to settle on the roadside tracts of land and cultivate food crops for sale in nearby towns and cities. Over time, the logging roads become farm to market roads, and deforestation increases along the roads. This dynamic may account for most of the losses of forests to croplands in sub-Saharan Africa. A similar dynamic has unfolded in the outer islands of Indonesia. After concession holders constructed networks of roads, pepper cultivators set up small farms along the roads.[16] While these patterns identify smallholder land clearing as the chief driver of forest losses in sub-Saharan Africa and Southeast Asia, the construction of networks of logging roads laid the groundwork for the subsequent land clearing. The access to the forests through the newly constructed roads in the concessions has also increased the pressure of hunters on wild game in the concessions.[17] For forest dwellers such as the Pygmies, the subsequent declines in wild game and bushmeat probably diminished their food security.

Ecocertification agreements between concession holders and the Forest Stewardship Council (FSC), the leading NGO that certifies environmentally responsible logging operations, have raised hopes that the spread of concession-based logging operations would not have deleterious environmental consequences. A comparison between FSC certified concessions and uncertified concessions in Gabon, Cameroon, and the Republic of Congo established that the roads in FSC certified concessions fragmented the forests just as much as did the roads in the uncertified concessions.[18] Sensitive to the deleterious consequence of logging roads, FSC staff amended their certification standards to limit road building in concessions and, in so doing, in the fragmentation of concession forests. FSC proved unable to enforce this measure, and a number of allied NGOs such as Greenpeace International, out of frustration, resigned from the FSC in 2018. The environmental coalition, in this instance, did not have sufficient power to manage the exploitation of the concessions in an environmentally responsible way. A recent state intervention in the Democratic Republic of the Congo to provide forest-dwelling Pygmies with rights to their ancestral forest homes holds out the promise of Indigenous-controlled forest concessions with a primary focus on harvesting wood in ways that maintain the integrity of old growth Congolese forests.[19]

China: A Strong State Induces Smallholders to Plant Trees

For its first forty years, the People's Republic of China pursued a policy of maximizing agricultural production in order to feed a rapidly growing urban population and to fund investment in a burgeoning industrial sector.[20] A Chinese variant of the Great Acceleration occurred after 1950, and it had negative consequences for rural landscapes. A succession of natural disasters in the last three decades of the twentieth century underscored the deteriorated state of the Chinese landscape and provided the impetus for changes in state policies about forests.

A crisis narrative emerged during the 1970s with the advance of the Gobi Desert into China's wheat-growing regions. To counter desertification and address impending shortages in the supply of wood, the government embarked on a vast tree planting campaign during the 1980s. High seedling mortality characterized this effort. The crisis narrative became more compelling after the 1993 "Black Wind" dust storm in northwest China and the 1998 floods along the lower Yangtze River. These events underscored the deteriorated condition of China's landscape and convinced party officials that China should embark on a major effort to increase forest cover.

The Black Wind dust storm underlined the large extent of wind eroded soils in northern China. The unprecedented Yangtze River flooding of 1998 killed tens of thousands of small farmers and demonstrated how the absence of trees in the upper reaches of the Yangtze watershed increased the vulnerability of people to flooding in the lower watershed. After the crises, party officials reformulated rural policies in an attempt to achieve three somewhat contradictory aims: to produce a stable stream of forest products, to improve the livelihoods of rural peoples, and to restore ecosystems. Given these goals, "dual purpose" forests became common.[21] The newly planted trees provided environmental services, and the sale of the trees or their fruit earned smallholders an income.

The party responded to the Yangtze flood with a six point reforestation program. Its best known component, the "Grain for Green" program, otherwise known as the "Returning Farmland to Forest Program' (RFFP), encouraged smallholders to forgo planting cereal crops on their fields and plant trees instead. While the trees growing in the former croplands matured, the participating farmers received an extra allotment of grain or income from the government. These plantings survived at a much higher rate than did the plantings from the 1980s. The RFFP represents the world's largest reforestation effort to date.[22]

Given the mammoth scale of Chinese society, the apex of government operates at a considerable remove from the everyday lives of citizens. In this structure, plans for change have a "high modernist authoritarian" orientation that makes their application in locales particularly problematic.[23] The state's initiatives applied across a wide range of places, but, at the same time, they frequently did not fit the circumstances of particular villages.

Given its remote bureaucratic origins, the implementation of the RFFP took unanticipated turns. The histories of the following four communities with the RFFP illustrate the recurring difficulties faced by communities participating in the program and the conditions that made the program successful. In each village, smallholders planted dual function species that would have ecologically restorative effects as well as commercial appeal to smallholder planters. The planted species, more often than not, did not take root as expected, but villagers and local public officials persisted with the program and eventually achieved a modicum of success. Consider four examples of this dynamic.

In Yunnan, a mountainous district near the Burmese border, the state forestry bureau provided Walnut trees to villagers, but the trees did poorly, rarely bearing fruit. Participating farmers replaced the government-provided walnuts with naturally regenerated walnut seedlings that the farmers dug up in the scrub growth surrounding the community. The choice of tracts of land to reforest led to disputes between local officials and RFFP staff. Local officials wanted the lands chosen for afforestation to include more farmers, in part because they wanted to increase the number of local beneficiaries from the program.

In northern Sichuan Province, every household with agricultural land participated in the RFFP. Farms averaged two-thirds of a hectare in size. Participating farmers retired about half of their lands and retained the rest of their land to grow crops. The government gave participating farmers a fast-growing holly tree whose leaves could be dried and sold as tea. The holly trees did not survive in large numbers, so villagers planted magnolias, valued for the medicinal properties of the bark after fifteen years of growth. The long delay in starting a stream of income from the magnolias entitled the villagers to compensation from RFFP well into the future.

In southern Sichuan Province, in an area inhabited by Tibetans, extensive deforestation had occurred during the previous thirty years. Some families began to reforest uncultivated lands on bare hilltops in 1998 with RFFP assistance. These efforts failed. Smallholders planted a fast-growing poplar tree (*Populus*) in the alluvial fan of a nearby river. Pigs and sheep consumed some of these seedlings, but others survived, in part because labor scarcities made it more difficult for some families to continue raising livestock.

Finally, on Hainan Island, the villages under study extended from the lowlands into the mountainous interior of the island. The inhabitants were mostly of Li or Miao ethnicity. The greater accessibility to overseas markets distinguishes the Hainan villages from the three sites discussed above. The participants in the RFFP differed as well. Smallholders only managed about 30 percent of the Hainan RFFP lands, some of which they converted into rubber plantations. Overseas investors managed the majority of the RFFP lands. They transformed uncultivated lands into eucalyptus and acacia plantations. The raw materials from these plantations went to mills owned by the outside investors. The economic returns from the plantations were so great that the subsidies provided through the RFFP did not make much of a difference in determining land uses in the Hainan villages. Given the genetic uniformity of the predominant planted species in Hainan and the other three communities, the environmental services of the RFFP seem limited to carbon sequestration and controls over runoff from rainstorms.[24]

The implementation of the RFFP program across these four disparate settings demonstrates a seriousness of purpose on the part of both state and local officials. At the same time, the remote authoritarian administration of the program increased the risks of program failure in villages. When failures occurred, farmers and local government officials tried again. As the case studies indicate, a high level of engagement by local officials in the RFFP's implementation frequently corrected for poor planning by the central state. The aggregate effect of this effort was considerable, at least as measured by recent increases in leaf area in satellite imagery.[25] Between 2000 and 2017, China accounted for disproportionate increases in leaf area across nations. An appreciable amount of the increase in China's leaf area probably came from increases in the extent of planted forests. A strong state, with shared goals among central government and local government officials, seemed essential to the success of the planting program. The shared goals among different cadres of officials also seems consistent with a state corporatist regime.

Important questions remain to be answered about the impact of the grain for green program on carbon budgets. Frequent harvesting of the reforesting acreage would call into question any claims that the program has created a carbon sink on RFFP lands. The tree farms would improve the water retention capacities of the RFFP lands, thereby reducing the threat of downstream flooding, but climate change suppression through carbon sequestration would diminish with the frequent harvesting of the trees and the associated greenhouse gas emissions. A long-term extension of the RFFP program of payments into the future could reduce, if not eliminate, the problem of frequent harvesting.

Vietnam: Tenure Reforms, Agricultural
Intensification, and Tree Farms

The Chinese experience with planted forests resonated with officials in neighboring countries, most obviously in Vietnam. Similar regimes ruled in both places in the late twentieth century. Both China and Vietnam had one party states with ambitious agendas for social transformation in rural areas. Not surprisingly, Vietnamese forest policy resembled Chinese policy. It too promoted the spread of small-scale forest plantations.

Like Chinese officials after the civil war, the Vietnamese central government after the Indochinese conflict used the nation's forests to earn foreign exchange for use in pursuit of a national agenda of urban industrialization. Governance in Vietnam in the immediate postwar period centered around collective organizations, either initiated or administered by the state. Agricultural cooperatives produced the staple crop, rice, most often in lowland paddy fields. State Forest Enterprises (SFE) controlled most of the forested highlands in the country. During the late 1970s and 1980s, the state sought sources of foreign exchange. To this end, the SFEs focused on cutting down old growth forests and shipping the timber or wood products overseas.[26] The SFEs also planted exotics such as eucalyptus and acacia in tree farms. By 1990, the postwar years of logging had degraded almost all of the accessible forests in the country. As cutover forests became more common, wood products became more expensive. A growing number of observers began to lament the degraded status and diminished extent of Vietnam's forests. A crisis narrative about forests became more common among party operatives.

Institutional transformations in the Vietnamese state shaped the course of the state's response to the crisis in the forests. The agricultural cooperatives and state enterprise system began to disintegrate during the 1980s when many observers, perhaps inspired by the reforms in China, began to question the legitimacy of these institutions in Vietnam. The decentralizing and liberalizing Doi Moy reforms transformed smallholder agriculture in the mid-1980s.[27] Families in rural communities occupied cooperative lands, both rice paddies and uncultivated lands in the adjacent hills.[28] The occupied lands became family farms. After 1993, families purchased forest lands, in effect privatizing them. The larger and wealthier households in rural communities acquired more of these paddy, old field, and forest lands than their poorer neighbors.[29] The example of tree planting provided by the early activities of the SFEs, the new government subsidies, and a sharp rise in the price of timber all spurred smallholders to plant trees.

Officials in the central government responded to the deteriorating conditions in the forests with a series of actions. In 1993, the national government banned the export of wood, restricted agricultural land uses in heavily forested regions, prohibited the felling of trees on sloped land, and authorized the sale of forest lands to private parties. In addition, the government set up the 327 program that provided subsidies to defray the costs of planting trees in degraded forests or on bare hillsides.[30] Government officials had also expanded the scale of this reforestation effort in 1987 with a commitment to establish or preserve an additional five million hectares of forest. Another program made credit available to smallholders at concessionary rates so that they could purchase seedlings for small tree farms. Institutional support came in other forms as well. The government built mills to process timber, hired extension agents in forestry, and established experiment stations with planted trees in woodlots.[31]

Market forces with geographic contours provided an additional impetus to create tree farms. Beginning in the early 1980s, the central government expanded the network of roads that linked rural districts with urban and coastal places. In so doing, the new infrastructure opened up economic opportunities for the new owners of paddy lands. These affluent smallholders took advantage of government subsidies for fertilizers to intensify agricultural activities on paddy lands close to roads and in some instances to increase the number of crops that they cultivated each year on these roadside lands. The intensified agricultural activities on lowland fields provided revenue to finance tree planting. At the same time, the additional labor demands in the lowlands increased labor scarcities in upland fields. The increased scarcity of labor made forestry a more appealing land use in these

upland settings given the relatively low labor demands of forest plantations.[32] Once again, agricultural intensification in one place seems to have
encouraged the expansion of forests in another place.

A similar dynamic unfolded after World War II in a quite different political economic context, in the mountains of northern Luzon in the Philippines.[33] Smallholders began growing vegetables for sale in urban markets. The growing labor demands in these lowland fields, coupled with
increased rural to urban migration, led smallholders to abandon shifting
cultivation on the hillsides. The abandonment of the hillside fields was
not complete. Households retained a claim to these lands by planting
them with pine trees. Hillsides that had been bare in the 1920s became
forest covered by the 1960s. Agricultural intensification by smallholders
in the valleys had promoted the expansion of small-scale forest plantations on the hillsides.

Upland landowners in Vietnam still had to decide whether they wanted
to plant trees, plant coffee, or allow for natural regeneration on the degraded
forests and bare hills. In general, landowners with the more accessible lands
chose to plant trees or coffee while landowners with more remote agricultural lands allowed natural regeneration to occur on their lands. The location of tree farms reflected these patterns. They concentrated in coastal
provinces and in areas with dense road networks such as places close to the
cities of Hanoi and Haiphong in northern Vietnam.[34] These tree farms were
small. In a study of tree plantations in central and northern Vietnam, the
plantations averaged 2.1 hectares in size.[35] In terms of their overall national
extent, tree farms increased from 1.0 to 3.5 million hectares in Vietnam
between 1990 and 2010.[36]

Central government initiatives, such as the National Five Million Hectare Reforestation Program, complemented these changes in infrastructure
and in subsidies from the 327 program.[37] The range of programs from the
visionary statements by the heads of national campaigns to the subsidies
for individual farmers in a devolved system of land use suggests a unity of
purpose across the forestry sector that is consistent with a corporatist
approach to policymaking. The long-term effectiveness of these rebranded,
climate change policies in Vietnam would depend, as in China, on the

frequency of harvests from the new tree farms. Lower frequencies of harvests would increase the contributions of the tree farms to climate stabilization.

Laos: Large Tree Farms in a Weak State

Laos became an independent country in 1954, and almost immediately its leaders were caught up in the civil strife surrounding the war to reunify Vietnam. Under these circumstances, the country's leaders placed a high priority on asserting national sovereignty over the more remote and forested upland portions of the country. To establish more control over outlying communities, the government initiated a land use planning program during the mid-1970s that resettled the hill tribes in the more accessible lowlands along the Mekong River.[38]

With most of its citizens practicing shifting cultivation, the Laotian state had few sources of tax revenue or foreign exchange. The state could, however, earn revenue by granting concessions to exploit natural resources to foreign investors. At the same time, bans on logging in China and Vietnam had encouraged timber companies to look overseas for business opportunities. They found them in Laos, where they negotiated large-scale forest concessions in the resettlement zones. The negotiations between the state and foreign investors typically did not include the recently resettled peoples living in communities in and around the concessions. The agreed upon forest plantations occupied large proportions of community land, in one instance, 600 hectares out of 1,800 hectares in a community. The conversion of these swidden-dominated landscapes into fast growing forest plantations of oil palm or rubber induced declines of 0 percent to 40 percent in aboveground carbon.[39]

The dedication of extensive areas to the production of wood products for overseas markets sharpened food insecurities among the resettled Laotian peoples by diminishing the extent of the land base that peasants could rely upon to produce their sustenance.[40] To provide a veneer of legitimacy to the reservation of these lowlands for tree farmers, the government zoned the lands in question and designated them as "degraded" forests that would be eligible for conversion into forest plantations. In practice, the degraded designation meant that local peoples had practiced shifting cultivation on these lands.

The new plantations produced rubber or rapidly growing trees such as eucalyptus and acacia. Over the course of several decades after 1990, the concession holders converted hundreds of thousands of hectares from degraded swidden lands into rubber plantations in Laos and other upland Southeast Asian regions.[41] The spread of rubber plantations across Laos and Cambodia occurred quickly, in part because the new forest plantations replicated what the foreign firms had already done in China and Vietnam. The managers of the new plantations sent the timber and rubber back to markets in China and Vietnam. In effect, these concessions amounted to land grabs by overseas firms with headquarters in nearby, Communist Party–dominated states.

Indonesia: A Weak State, Planter–Peasant Ties, and Ecocertification

The outer islands of Indonesia contain exceptionally biodiverse, carbon rich forests with large dipterocarp trees and extensive deposits of peat. A humid climate sustains these biological communities. The profusion of life in these forests also makes them attractive settings for intensive agriculture. The humidity and the year-round growing seasons raise the productivity of agricultural enterprises. The prospect of large harvests has, in turn, attracted investors from overseas who sought concessions to log the old growth forests and then establish plantations in the concessions.[42] Reforms after the fall of the Suharto regime in 1998 decentralized control over the concessions, but the reforms did not eliminate the lax controls and influence peddling that governed the exploitation of the concessions. By 2016, plantations occupied 13 percent of the land area in Indonesia.[43]

The Great Acceleration, in the form of rapid economic growth and urbanization in China and India during the late twentieth century, fueled demand for cooking oil exports from Indonesia.[44] The growing demand induced sustained increases in global prices for palm oil during the 1990s and 2000s. Malaysian enterprises, already experienced in the production of oil palm on large plantations, applied their capital and skills to the development of Indonesian oil palm plantations in the outer islands (see figure 5.1). Between 1990 and 2005, planters converted about three million hectares from old growth forests into oil palm plantations. Overall, the extent of planted oil palms in Indonesia increased from 1.1 million hectares in 1990 to 11 million hectares in 2016.[45] By 2010, Indonesia produced more than one-half of the world's total harvest of oil palms.[46]

The decentralization of forest concession management in the late 1990s facilitated the growth of smallholder participation in the expansion of the plantations. By 2016, small parcels contained 36.3 percent of all the area planted in oil palm in Indonesia.[47] Smallholders did this work in association with large-scale oil palm companies. The companies created oil palm plantations for smallholders. In return for creating a three-hectare plantation on a smallholder's land, the company would receive unused land from the smallholder, as much as seven hectares in extent. The companies then integrated these lands into their own plantations. Fruit from both the smallholding and the company holding would get processed at the company's oil palm mill.[48] The profitability of the oil palm business reduced poverty among

FIGURE 5.1 Oil palm plantation, 2020, Kalimantan, Indonesia. Photo courtesy of Arrowsmith Films.

smallholders with ties to the larger plantations. Oil palms in plantations, given their large stature, did sequester carbon at high rates (6.1 t C ha^{-1} yr^{-1}).[49]

The environmental price for these economic and ecological gains has been fearsome. In the most notorious instances, palm oil agribusinesses drained and dried out peat lands that, during droughts, then caught fire and emitted very large volumes of greenhouse gases. These initial losses of biodiversity and stored carbon continued as the plantations aged. A study that compared adjacent planted forests and agriculture–forest mosaics in Indonesia found that land users cleared the planted forests at higher rates than they cleared lands in the agriculture–forest mosaics.[50] With each clearing comes additional emissions, so creating plantations with short cycles of planting would not meet carbon sequestering goals.

The plans for carbon sequestration through tree farms presumes a quite different pattern of land use: forest plantations where people plant trees and leave them alone for forty years. The pattern that prevailed in Indonesia during the 1990–2015 period reflects a forestry sector in which widespread illegal logging and mill construction depressed prices for wood. The low prices encouraged loggers to liquidate their holdings through early harvests and invest in other industries. A weak state, unable to manage its forest concessions, may explain the rapid development of tree farms, the frequent replantings, and the associated surge in greenhouse gas emissions from these farms during the 1990–2015 period.

The contributions of smallholders and large plantations to oil palm–related deforestation differed considerably. Oil palm production required considerable infrastructure. The fruit had to be processed in a mill relatively quickly after harvesting. Quick processing requires proximity to a mill and a network of roads that make it possible to transport the fruit to the mill. Smallholders did not have the funds to build the infrastructure. For this reason, they committed to oil palm production only in places where companies had already put the infrastructure in place. In these locales, large expanses of forests had already been cleared, so smallholders usually cleared little additional land when they planted oil palm on their farms. The major losses of old growth forests occurred earlier in time when workers for large investors moved into largely forested zones, cleared forests, planted oil palms, constructed mills to process the fruit, and built roads to get the fruit to the processing mills. Given these patterns, large-scale oil palm planters have accounted, directly and indirectly, for most of the recent destruction of old growth forests in the outer islands of Indonesia.

How then can institutions salvage the carbon sequestering capacity of an oil palm plantation without incurring additional environmental damages? Studies of the above ground carbon content of oil palm plantations have yielded a wide range of results, from 33 t C^{-1} to 91 t C^{-1}.[51] These variations in aboveground carbon content may stem from the presence or absence of sustainable practices in the cultivation of oil palm. New non-profit organizations have encouraged the adoption of more sustainable practices. The Roundtable on Sustainable Palm Oil (RSPO) is the best known of these initiatives. It began in 2002 as an initiative of large businesses that traded palm oil or palm oil-based products. RSPO personnel identified a set of environmentally benign practices that, if followed, would reduce environmental damages from the cultivation of oil palms. If a producer compiled with these practices, she or he would earn certification as a sustainable producer of palm oil. Certification as "sustainable" would attract environmentally concerned consumers to that brand of palm oil and earn the producer higher prices for her/his product.

By 2018, approximately one-fifth of the world's cultivated palm oil lands were so certified. However well intentioned, the RSPO certification scheme has not led to substantial differences in the cultivation of oil palms in the one study that compared certified and noncertified plantations.[52] When initially established in 2004, RSPO's founders required that plantations, to be certified, could not have deforested primary forests after the end of 2005. The initial roster of participating plantations contained large numbers of recently deforested landholdings with few, if any, tracts of primary forest within their boundaries. Prohibitions on forest destruction did not constrain the managers of these enterprises because they had already cleared almost all of the forest from their farms. In this respect, the RSPO rules had little if any impact on deforestation in the oil palm–producing regions. Most of the oil palm plantations who chose to join RSPO had little forest left, so their environmental profiles did not look very different in terms of conserved forests than did the environmental profiles of neighboring plantations that did not join RSPO.[53]

The Southern United States: Urbanization, Forest Plantations, and Carbon Sequestration

The previous chapter described how cotton croplands in the American South deteriorated by the mid–twentieth century to the point where sharecroppers and landowners abandoned these lands. Regenerating secondary forests took their place during the 1950s and the 1960s. Beginning in the 1970s, the composition of the remaining and regenerating forests in the southeastern United States began to change. The earlier cycle of land abandonment and native forest regrowth persisted in a few places with rugged terrain where smallholder agriculture had predominated, such as southeastern Oklahoma. Outmigration to cities had continued throughout this period, to the point where few people remained on the marginal lands. Closer to cities, real estate developers began to carve up naturally regenerated forests into building lots for subdivisions of single-family homes, so the overall extent of naturally regenerated forests declined.[54]

While naturally regenerating secondary forests declined in extent, planted forest area increased across the southeastern United States. Planted forests increased from 8.5 percent of all forests in the South in the mid-1970s to 23.0 percent of all forests in the South in 2015.[55] Foresters planted loblolly pine (*Pinus taeda*), a tree native to the region, in many instances. Overall, forest cover continued to increase in the southeastern United States, by 6.3 percent between the mid-1970s and 2015. It increased during this period, but only because planted tree cover increased tremendously, by 188 percent between the 1970s and 2015. Naturally regenerated forests declined by 10.5 percent in extent during the same period.

An increase in harvested area coincided with the increase in planted area. The harvested area in the South doubled between 1986 and 2011.[56] The coincidence of the two trends is not accidental. Timber product removals were much higher in counties with tree farms than they were in counties where naturally regenerated forests predominated.[57] This pattern of timber removals in the South repeats the pattern that Sean Sloan and his collaborators found for planted forests in Indonesia.[58] The rapid rate of harvesting in the planted forests of the South has conceivably affected the overall carbon budget for the southern landscape. Observers might expect the southern forests to be net carbon sinks given their overall increase in extent since World War II. This condition has changed over time. As the proportion of planted forests, with their rapid rates of harvest, increased over time, the southern forest slipped from being a sink during the 1980s, sequestering more carbon than it gave off, to becoming a source after 2000, emitting more carbon dioxide than it sequestered.

The consumer demand that sustained the high rate of harvesting in planted forests came from two general sources. Paper accounted for 40 percent of all of the timber, and residential construction in metropolitan areas accounted for another 28 percent of the harvested timber.[59] Growth in the size of homes during the postwar era most likely contributed to the burgeoning demand for wood products. The geography of planted trees in the South becomes intelligible in these terms. Tracts of planted trees concentrated in the Piedmont region of the South, close to the rapidly growing metropolitan areas of Birmingham, Atlanta, and Charlotte. The proximity of the mills to metropolitan areas reduced the transportation costs of getting the timber into the hands of the home builders.

Most recently, the industry began to create wood pellets in the mills for export overseas as biofuels. The largest single source of energy for Great Britain now comes from the biomass burning Drax power station in northern England. It generates power by burning wood pellets imported from pine plantations in the southeastern United States.[60] The British reliance on biomass burning as a source of energy stems from a commitment in 2009 by the European Union to renewable energy as part of a larger program to reduce their reliance on fossil fuels.

Government interventions, dating from the New Deal era and reformed during the 1980s, have spurred recent growth in the southern forests. The federally funded Conservation Reserve Program (CRP) pays the owners of both plantations and naturally regenerated secondary forests for conservation enhancements such as increases in tree cover on their properties. The beneficiaries of the CRP vary dramatically from county to county across the South. In some regions, smallholder forest owners, sometimes absentee, predominate, and they benefit from the CRP. They sell their timber to pulp and paper mills owned by large companies. In other settings, companies own both the land and the mills. They too benefit from the CRP. Employees of the company log the forests and process the timber in mills owned by the company. The concentration of wealth, especially in the second pattern, runs along racial lines, with whites holding much more wealth than African Americans.

In an era of climate change, the CRP would provide the institutional basis for an expansion of its core mission to bolster environmental services coming from the land, such as carbon sequestration. The CRP would engage both the owners of small forest plantations and the companies with large plantations. To date, discussions about the likely shape of a program of payments to enhance the carbon sequestering capacity of southern forests have not begun. To reestablish the southern forests as a carbon sink, this type of program would have to lower the frequency with which owners chop down the trees on their lands.

Chile: A Plantation-Driven Forest Expansion

Chile's forests exist in a biogeographically isolated region to the west of the Andes, to the east of the Pacific Ocean, to the south of the Atacama Desert, and to the north of the windswept and mostly treeless Patagonia. The isolation may explain the high rates of biodiversity and endemism in Chile's native forests. When people have deforested and degraded these lands, correspondingly large losses in biodiversity have most likely occurred.

Successive waves of forest cover change have washed over the Chilean landscape since the 1600s. Indigenous peoples had cleared close to a million hectares for crops prior to the arrival of the first Europeans. Armed conflict and epidemics that coincided with the arrival of the Europeans led to large losses of population, and much of the Native American–cultivated land reverted to forest. Immigration by European settlers during the second half of the nineteenth century led to further land clearing for agriculture and accelerated logging to meet the demands for urban construction materials. Import substituting industrialization after World War II raised the prices of agricultural inputs such as fertilizers from overseas and eroded the profitability of crop exports, so agriculture contracted, and forests expanded between 1950 and 1980. Beginning in the 1970s, the drivers of forest expansion shifted toward planted trees. In 1974, the Chilean state established a program to subsidize the creation of forest plantations. Trade liberalization after 1980 expedited the expansion of wood exports from the recently established forest plantations.

Table 5.1 charts these large-scale changes in Chilean forest cover.[61] Different measurement procedures underlay each of these estimates, even the most recent figures for the 1980–2020 time period, so there is an air of misplaced precision in the numbers in the table. Nonetheless, the aggregated data do suggest an overall increase in forest cover after 1980.

A change in the composition of Chilean forests occurred during this period. Planted forests, featuring pine (*Pinus radiata*) and eucalyptus (*Eucalyptus spp*) expanded rapidly, oftentimes at the expense of native forests. In central Chile between 1986 and 2011, forest plantations increased by 100.52 percent in extent while native forests declined by 12.86 percent in extent.[62] The forest plantations expanded onto former agricultural lands (45.1 percent of converted lands), into shrublands (31 percent of converted lands), and into native forests (22.3 percent of converted lands).[63] While the expansion of forest plantations did lead to further losses of native forests, forest plantations, by 2010, produced 95 percent of Chile's annual harvest of wood from only 15 percent of its forested lands. Some analysts have argued that the wood production from the plantations has reduced the

TABLE 5.1 The Extent of Chilean Forests: Historical Fluctuations

Date	Estimated forests: Millions of hectares	Source
Before 1500	30	Lara et al. 2012
1850	24–29	Otero 2006
1913	11–16	Albert 1913
1946	10.9	Haig et al. 1946
1960s and 1970s	14	INFOR 1986
2020	18.2	FAO 2020

economic pressure on native forests in Chile. The growth in planted trees may account for recent declines in the rate of deforestation in native forests in central Chile from .6 percent per annum from 1986 to 2001 to .1 percent per annum from 2001 to 2011.

The subsidies for the expansion of forest plantations in Chile did not lead to an increase in carbon sequestration in the country, primarily because the native forests destroyed by forest plantation expansion contained so much stored carbon that was lost through emissions. The native forests in the southern forest zone contained 146 tons of aboveground carbon per hectare compared with 37 tons of aboveground carbon in forest plantations.[64]

Large Chilean timber companies have played central roles in both the international timber trade and in the expansion of forest plantations in Chile, so these companies became the targets of moratoria, certification schemes, and other campaigns to reduce deforestation, beginning during the 1990s. Analysts refer to these compacts and campaigns, collectively, as "non-state, market driven governance."[65] During the 1986–2011 period, the Chilean timber companies participated in these governance regimes with three NGO certifiers: the Forest Stewardship Council (FSC), the Sistema Chileno de Certificación de Manejo Forestal Sustentable (CERTFOR), and the Joint Solutions Project (JSP). The FSC, as noted in the Congolese case study, is an internationally recognized third party certifier of sustainable forest practices. CERTFOR, an association of large timber producers in Chile, established a set of principles to govern the management of plantation forests in Chile, and it certifies plantations when they comply with CERTFOR's principles of practice. JSP, like FSC, is a third-party certifier of sustainable forestry practices. JSP arrived at its principles of practice through meetings, arranged by Home Depot, between two large Chilean timber companies and environmental advocates in the United States.[66] All three of these organizations would not certify tree farms that converted native forests into tree farms during the period of observation. Among the three certifiers, the FSC had the reputation of having the most environmentally restrictive standards of management because it applied its standards retroactively. It refused to certify tree farms that had deforested native forests after 1994. FSC participants had lower historical deforestation rates than did participants in the JSP and CERTFOR compacts.

Tree farms that participated in these agreements did reduce the deforestation of old growth forests in their operations by anywhere from 3 percent to 22 percent compared with tree farms that did not participate in any of the three certification processes. While the increments in forest protection observed in the participating tree farms were not large, they did occur on a selected set of properties where the economic pressure to deforest may have been greater than normal. The participating tree farms tended to be larger in size than nonparticipants and located in more accessible locations where, presumably, the pressures to deforest the remaining native forests would have been greater. Participating tree farms also tended to be located on prime agricultural lands. There was little evidence of leakage effects where the adoption of non-state, market-driven certification schemes led to more destruction of naturally regenerating forests in neighboring, nonparticipating territories.[67]

The forest plantation companies varied in the degree of conflict they experienced in their contacts with environmental NGOs. FSC participating companies worked collaboratively with NGO representatives to come up with FSC compliant protocols for forest management. These companies also reduced their clearing of native forests to a greater degree than did companies who participated in other non-state, market-driven certification compacts.[68] The superiority of the more collaborative approaches in reducing environmental damages from forest plantations supports arguments about the value of corporatist approaches to environmental governance. For example, the organization of interests into sectoral associations usually reduces the amount of class conflict within a sector or in this instance between activists in environmental NGOs and elites in the timber companies.

Increased carbon sequestration through an expansion in planted forests figures centrally in Chile's plan for national contributions to global greenhouse gas–emissions abatement. To this end, the Chilean Wood Corporation, a trade association of forest plantation owners, has proposed the creation of an additional two million hectares of forest plantations in Chile. The new plantations would take up 2.6 percent of Chile's land area. In addition, the federal government would allocate an additional $37 million per year to help defray the costs of establishing the new plantations.[69]

The Expansion of Tree Farms: General Patterns

Crisis Narratives, Strong States, and Tree Farms

Crisis narratives now characterize most discussions of policy alternatives to combat climate change. An "all hands on deck" mentality marks discussions. Policy options that promise some climate stabilization seem palatable, even though they may exacerbate other problems. Tree farms represent one of these partial solutions. In circumstances where a tree farm replaces an old growth forest, the biodiversity losses and greenhouse gas emissions from the initial forest destruction are so large that it becomes impossible to argue on environmental grounds for the creation of tree farms. In those instances where planted trees replace bare ground, croplands, or degraded forests, the carbon they sequester above and below ground may compensate for the losses of biodiversity that occur when a monocrop of trees replaces already degraded plant communities.

With the climate crisis and these partial solutions in mind, planners have assigned tree farms a prominent place in mitigation strategies. Dozens of nations at the Paris Conference of Parties in 2015 pledged to achieve emissions reductions through expanded carbon sequestration in planted trees.[70] The plans submitted by the forty-three nations participating in the Bonn Challenge confirm the importance of tree farms in the proposed forest restorations. As noted in an earlier description of the plans, tree farms would make up more of the reforested area than either naturally regenerated forests or agroforests.[71] In countries with largely deforested and degraded landscapes such as Haiti, tree farms may represent one of the few feasible ways to rehabilitate a landscape.

Starting a tree plantation requires that the planters mobilize a labor force to plant and care for the trees. When governments do the planting, their capacity to plant trees and care for them becomes an important consideration. Strong states, with ample capacity to mobilize people to accomplish collective ends, could carry out massive tree planting campaigns.[72] Crisis narratives, in this instance about climate change, strengthen states by alarming citizens and encouraging them to contribute to the common cause that solves the problem. Weak states, with little capacity to mobilize people for collective action, would find it difficult to accomplish goals that require citizens to plant and manage thousands of hectares of trees.

The strength of states varies across forms of government. South Korea, at the time an autocracy, demonstrated considerable state strength during the 1960s and 1970s when it mobilized citizens to replant forests devastated by successive wars. In the United States, long a democracy, Franklin Delano Roosevelt created the Civilian Conservation Corps, which planted hundreds of thousands of hectares of forest during the 1930s.[73] The Depression era United States exhibited the essential characteristics of a strong state. As these historical examples suggest, the strength of states fluctuates over time. The Paris agreement presumes that the climate crisis will increase the strength of states. The newly strengthened states would, in turn, facilitate the spread of state-sponsored tree farms.

As the case studies of Indonesia, the United States, and Chile indicate (see table 5.2), the spread of planted forests has often been part of a larger, preexisting, market-driven pattern of deforestation. Given the economic calculations of new plantation owners in this context, the newly planted forests did not last long. As soon as the planted trees were large enough to sell in an urban market, the owners harvested them. These forest plantations became part of a treadmill of production in which landowners tried to earn as much as they could through repeated harvests of trees at frequent intervals.[74] The frequent harvesting nullified the carbon sequestering effects of tree planting.

TABLE 5.2 The Political Economy of Forest Concessions and Plantations

Nation	Crisis narrative?	State strength	Overseas investors?	Market-driven?	Smallholder focused?	Food security diminished?
Congo	No	Low	Yes	Yes	No	Yes
China	Yes	High	No	No	Yes	No
Vietnam	Yes	High	No	Yes	Yes	No
Laos	No	Low	Yes	Yes	No	Yes
Indonesia	No	Low	Yes	Yes	Mixed	No
United States	No	Medium	No	Yes	No	No
Chile	No	Medium	No	Yes	No	No

Sources: Case studies in chapter; see notes.

As suggested in the case study of forest plantation expansion in Laos, the emphasis on accessible locations for forest plantations has had negative implications for the food security of smallholders living in adjacent areas. Forest planters usually chose croplands as sites for forest plantations because these lands tended to be close to urban centers with their concentrations of consumers. Remote sensing analyses of the land use conversions accompanying the expansion of forest plantations in northern Mozambique between 2001 and 2017 indicate that 70 percent of the converted lands had been croplands before landowners converted them into tree plantations.[75] Under these circumstances, the loss of croplands to forest plantations reduced local sources of foodstuffs and imperiled the food security of impoverished local peoples.

Trajectories of Change Among Large-scale and Small-scale Planters

Tree plantations vary tremendously in size from region to region in the Global South. In the densely populated agricultural districts of East Africa, South Asia, and East Asia, plantations may average no more than two hectares in extent. Small-scale kitchen or domestic tree farms have proliferated in these places. Unlike forest plantations which have one or two species of trees planted in evenly spaced rows, kitchen forests have clusters of planted trees of various species. Smallholders planted trees out of the back doors of their houses, sometimes as part of a tree planting campaign. While small in size, these woodlots and small plantations, in the aggregate, have global significance. In East Asian countries with little or no old growth forest and large rural populations, the scarcity of wood has incentivized the planting of fast-growing trees in small-scale forest plantations. Tree planting initiatives spread throughout the East Asian region after 1990.[76] By 2020, South and East Asia contained 46.2 percent of the world's land area in tree plantations.[77]

In stark contrast, individual tree plantations in Indonesia, Chile, and the southern United States often contained more than one thousand hectares of trees. If large-scale planters had a mill on the premises for processing products, they often built ties to nearby smallholders who agreed to send their harvests to the mill for processing. The number of small-scale oil palm producers with these kinds of links to large-scale planters grew rapidly in Indonesia after 2000. In a similar dynamic that unfolded after 1980, wood-lot owners in the southeastern United States contracted with the nearby sawmills of large timber companies to process their timber.[78]

Smallholder plantations have also become a focus for agricultural inten-sification efforts. While small-scale oil palm planters obtain from 1 to 3.7 tons of palm oil per hectare in sub-Saharan Africa and Southeast Asia, sci-entists in experiment stations have obtained yields as high as 10 tons per hectare.[79] Research and outreach could close this yield gap between exper-imental farms and smallholder plantations. The associated increase in pro-ductivity per hectare could in turn reduce the extent and the damaging environmental impact of commercially oriented tree farms. The use of improved planting material, better timed use of fertilizers, and the recy-cling of waste from the oil palm mills all promise to reduce the amounts of environmental damage by reducing the extent of any further expansion of oil palm plantations.[80] These examples of ecological modernization would reduce the emissions per unit of palm oil produced.[81] Conceivably, by concentrating production on smaller tracts of land, these changes could free up additional tracts of land for forest restoration.[82] Of course, the pos-sibility exists of a contrary effect, in which the increased productivity per hectare spurs large and small plantation owners to expand their planting of oil palms.

PES and REDD+ programs in smallholder districts would have markedly different distributional effects than they do in places where large-scale plantations predominate. If the distribution of forest plantations in small-holder dominated districts follows the relatively equitable distribution of land in these regions, then payments for environmental services should show, unlike in places where land distribution is concentrated, an equita-ble distribution among the participating landowners. The implementation of PES and REDD+ programs in these contexts would have important sym-bolic consequences. It would provide tangible evidence to all members of the global public that even the rural poor have a meaningful role to play in the crusade to prevent catastrophic climate change.

Plantation Understories: A Site for Sustainable Development

The negative impacts of most tree farms on the biodiversity of landscapes probably diminishes over time. As the planted trees establish a canopy, they create shade and preserve humidity. Under these conditions, a species rich understory of tree seedlings frequently emerges (see figure 5.2). The plant growth attracts animals who disperse seeds and, in so doing, improve the biodiversity of the emerging understory.[83] Absent frequent harvesting, the understories of tree plantations begin to resemble naturally regenerated secondary forests in a locale. Given these similarities, the rationales for promoting secondary forests as a natural climate solution begin to apply to forest plantations. If the creation of a plantation brings with it a commitment to allow old growth stands of trees with biodiverse understories to emerge, then these plantations could well contribute to climate stabilization and other environmental services.

In southern Brazil, land change scientists and activists have begun to designate tracts of planted forests with managed understories as sites for

FIGURE 5.2 Forest plantation in Uganda with a growing understory. Photo courtesy of the UN Food and Agriculture Organization.

ecocertified succession. In experiments, landowners have planted fast-growing, carbon-absorbing exotics such as eucalyptus on tracts of land. In as little as five years, the exotics grew into an overstory that had sufficient value to harvest and sell. The sale of the fast-growing exotics has provided landowners with an economic incentive to pursue a transition in types of forests, from a plantation of exotics to a naturally regenerated secondary forest. The transition takes place as follows. At the same time that landowners plant the exotic trees, they allow native vegetation to grow into the understory. In some instances, the landowners plant native species in the understory. With the passage of time and the beginning of the harvest of the exotics in the overstory, the native secondary growth in the understory gradually becomes the dominant assemblage of plants in the forest, bringing with it increased biodiversity and rapid rates of aboveground and belowground carbon sequestration.[84]

Planted forests, if harvested infrequently, do represent an opportunity for ecologists. The understories of tree farms, depending on how foresters manage them, do become biodiverse. After fifty years of growth adjacent to a naturally regenerated secondary forest in Puerto Rico, the species richness of plantation understories approached the species richness of understories in comparable, naturally regenerated secondary forests.[85] More generally, plantation forests and their understories could, conceivably, be managed in ways that would both promote the establishment of carbon sequestering forests and bolster the biodiversity of the plant communities that emerge in and around the new forests.

Concessions and Plantations: From Engines of Destruction to Devices for Restoration?

Beginning during the postwar economic recovery in the mid–twentieth century, farmers, loggers, and planters destroyed old growth forests, and planted large tree farms in a wide range of places, especially in insular Southeast Asia. When they did so, they caused big losses in biodiversity and triggered large-scale emissions of greenhouse gases. Fires set by people clearing land exacerbated the ecological devastation that accompanied the creation of the forest plantations. The predominant monocultures in the plantations suppressed biodiversity in the emerging plantation landscapes.

Other conversions to forest plantations suggest more environmentally benign outcomes. When smallholders in China and Vietnam converted degraded agricultural lands into small plantation forests, the new forests, compared to previous land uses, sequestered increased amounts of carbon in many settings. Large scale conversions of grasslands in southern Brazil into new, highly productive eucalyptus plantations increased carbon sequestration on these lands. The yields from the eucalyptus in these projects were so high that the participating timber companies agreed to allow one-half of the project's grasslands to revert to scrub growth and naturally regenerating forest.[86] In its overall contours, this agreement resembles the real estate development—conservation deals negotiated by developers and conservation advocates in the affluent suburbs of the Global North (see chapter 3). In return for accepting an intensified zone of land use, in this case a forest plantation, conservation advocates got a protected area with naturally regenerating forests.

Beginning in the early 2000s, land change scientists have tried to create environmentally friendly certification schemes that would increase the environmental services from plantations. The initial impacts of these certification schemes have been disappointing. They have not prevented forest fragmentation in the Congo, and they did not slow down rates of deforestation in Indonesia.[87] More comprehensive certification schemes, with attention to the ways planters use peatlands and their record for fire prevention might increase the impact of certification on land use in the outer islands of Indonesia. More robust certification effects would be more likely if the price for certified palm oil were to increase. In this regard, it weakens the certifiers when many palm oil purchasers, concentrated in poorer countries such as India, refuse to pay higher prices for sustainably produced oil palm. These considerations suggest that creating large flows of sustainably produced palm oil from plantation forests faces daunting challenges.

The strategy of concentrating production in some places in order to spare other places from production applies to plantations. For example, the large yield gaps in production per hectare between oil palms and other vegetable oil crops offer an opportunity for reducing the damaging impacts of agricultural expansion on tropical forests. Oil palms produce double the volume of production per hectare of other vegetable oil crops such as peanuts, soybeans, coconuts, rapeseeds, olives, and maize.[88] If oil palm were to replace these other vegetable oils, the enhanced production per hectare of oil palm could, *ceterus paribus*, free up substantial amounts of cropland currently devoted to the cultivation of the other vegetable oils. Alternatively, the substitution of palm oil for other vegetable oils could make the cultivation of oil palms even more profitable and, in a rebound effect, induce growers to expand the extent of their oil palm plantations at the expense of existing rainforests.[89]

The gains in carbon sequestration in these cases could, however, be overstated if analysts presume that the forest plantations will endure, locking up carbon for a half-century, if not longer. Currently, harvests on plantations occur much more frequently.[90] Harvesting rates will have to slow dramatically if forest plantations are to become carbon sinks. It will take strong states, organized in a societal corporatist form, to accomplish this end. To be credible, these projections need to acknowledge the different political dynamics that have characterized the tree farm sectors in the case studies presented here. China and Vietnam have constructed state corporatist regimes in which political elites and smallholders have displayed a unity of purpose. Chile displays a societal corporatism in which timber companies, legislators, and NGO representatives have begun to discuss a common agenda for expanding tree farms in order to increase the nation's volume of carbon sequestration. Laos, Indonesia, and the United States seem further removed from attaining coherent, globally significant, and environmentally defensible tree planting efforts.

6

• • • •

Agroforests I

THE SPREAD OF SILVOPASTURES

Grasslands with sparse stands of trees, often referred to as savannas, occupy 36 percent of Earth's land surface. Savannas sequester carbon, especially in circumstances where cultivators leave them undisturbed.[1] Trees, if provided with caretakers, do grow in some of these settings. For this reason, some analysts have come to regard tree–grass landscapes as potential sites for carbon sequestration through afforestation.[2] Because these landscapes have historically contained so few trees, the creation of forests or silvopastures in these settings would sequester additional amounts of carbon and contribute to natural climate solutions.[3] The arid climates in many of these places would limit the growth of trees, but the large potential extent of these forests leaves open the possibility that programs to increase tree cover in these settings could make a significant contribution to a natural climate solution.

The addition of trees to grass landscapes usually takes the form of silvopastures that integrate trees into plots of land that people manage for livestock. These silvopastures represent a form of agroforestry. Pastoral landscapes have expanded rapidly during since the 1970s as cattle ranching became the primary driver of deforestation in the Amazon basin.[4] Emerging silvopastures vary considerably, from pastures with just a few trees per hectare to pastures with three or four hundred trees per hectare.

The advent of remote sensing technologies that can distinguish land-scapes with sparse tree cover from pastures and row crops have made it possible to measure the changing extent of agroforestry both regionally and worldwide. Since about 2010, researchers have begun to use medium reso-lution MODIS satellite imagery to measure changes in the canopy, the pro-portion of the sky covered by trees in a field. By aggregating these observa-tions for a region, it has become possible to compare the changing extent of tree cover in agricultural lands across different regions of the world.[5] Table 6.1 reports these trends.

The proportion of agricultural lands with significant tree cover has increased since 2000. The global proportion of agricultural lands with greater than 10 percent tree canopies increased from 40 percent to 43 per-cent over the ten year period between 2000 and 2010.[6] Several patterns stand out in these gains in forest cover. First, precipitation promotes the spontaneous germination of tree seedlings in agricultural land. The three regions with the highest annual average precipitation—Southeast Asia, Central America, and South America—had the most extensive tree cover

TABLE 6.1 Changes in the Extent of Agroforestry (Percentage of all Agricultural Lands) by Region, 2000–2010

	2000–2002	2008–2010	Change over time
North America	40.4	42.2	+1.8
Central America	94.5	96.1	+1.6
South America	53.0	65.6	+12.6
Europe	45.0	45.0	0.0
North Africa and West Asia	10.1	11.0	+0.9
Sub-Saharan Africa	28.6	30.5	+1.9
North and Central Asia	28.2	25.3	−2.9
South Asia	21.0	27.7	+6.7
Southeast Asia	76.9	79.6	+2.7
East Asia	42.6	47.5	+4.9
Oceania	30.3	33.3	+3.0

Note: Tree cover on agricultural lands (percentage of agricultural lands with greater than 10 percent tree cover)

Source: Robert Zomer et al., "Trees on Farms: An Update and Reanalysis of Agroforestry's Global Extent and Socio-Ecological Characteristics" (Bogor, Indonesia: World Agroforestry Centre (ICRAF) Southeast Asia Regional Program, 2014).

in agricultural lands. All three regions saw significant increases between 2000 and 2010 in the proportions of agricultural land with large numbers of trees. Second, growth in agroforested landscapes during the 2000–2010 period extended beyond sparsely treed landscapes. In Central America, the proportion of agricultural land with tree cover exceeding 20 percent of the land area increased from 70.8 percent in 2000–2002 to 79 percent in 2008–2010. Agricultural lands with tree canopies exceeding 30 percent of fields increased from 47.2 percent to 54.8 percent of all agricultural landscapes in Central America from 2000–2002 to 2008–2010.[7]

Particular types of agricultural land uses—pastures compared, for example, with row crops—seem more likely to evolve into agroforested landscapes. Plowing in row cropped landscapes eliminates tree seedlings. In contrast, the upkeep of pastures often does not entail the uprooting of tree seedlings, so these fields may over time become populated with trees. Cultural differences, traceable in some instances to the colonial era, may account for some of the differences between regions in the proportions of agroforests, croplands, and pastures. For example, Spanish colonizers in Latin America had a long-standing cultural preference for raising cattle that they made manifest by establishing extensive cattle ranches in the neotropics.[8] Pastures often become silvopastures, in effect agroforests, when tree seedlings from adjacent forests spontaneously germinated in pastures and landowners allowed them to grow.

In both arid and humid settings, the emergence of some trees in pastures often intensifies land uses. Provided that shade tolerant pasture grasses predominate in a place, an additional economic activity, the care for and harvesting of trees, now occurs without a commensurate decline in cattle on a tract of land. The sale of trees and associated wood products such as fuelwood diversifies the smallholders' incomes. They no longer depend exclusively on income from livestock. The sale of wood products may have important poverty alleviating effects in arid settings where smallholders' incomes are very low.[9]

Some case studies in the growth of silvopastoral landscapes follow. Different dynamics drove each expansion. Viewed comparatively, these dynamics suggest the conditions and circumstances that have promoted the spread of trees across pastoral landscapes.

Spontaneous Silvopastoral Landscapes
in the Upper Amazon Basin

Silvopastoral landscapes have begun to emerge since the early 2000s in recently deforested regions of the Upper Amazon basin in Ecuador. Rough terrain from the Andes meets the Amazon plain in these locales. Orographic rainfall drenches the region, with anywhere between 2,000 mm and 4,000 mm falling annually without notable seasonal variations. Burning was out of the question in this wet environment, so shifting cultivation in the region took the form of "slash and rot" rather than "slash and burn." For most of the twentieth century, the practical difficulties of constructing roads in rough terrain defeated construction crews and a fiscally strapped national government. The lack of road access slowed attempts by colonists to occupy, claim, and clear forests in the region. The isolation of the region began to diminish during the 1960s when the Inter-American Development Bank agreed to finance the construction of roads from the Andes down into the Amazon basin.[10]

Anticipating eased access to the region, colonists and communities of Shuar, the longtime Indigenous inhabitants of the region, claimed tracts of forest in corridors along the proposed routes for the roads. Neither the Shuar nor the mestizos had appreciable amounts of wealth. They established small, ethnically segregated settlements and carved small farms out of the forest ranging from thirty to seventy hectares in size. To strengthen their claims to land, colonists and the Shuar cleared forests and planted pasture grasses on the cleared land. The colonists clear cut their fields, leaving few if any trees standing. They did so out of fears that thunderstorms could topple the remaining trees and kill the cattle grazing at the base of the trees. Farmers did not fence their fields. They tethered their cows to trees or to bunches of grass. Each morning and afternoon, a member of the family would move each cow into mature grass. During the next twelve hours, the cow would eat the forage within reach. Over the course of a week, a line of tethered cows would move across a pasture. The person who "changes the cattle" also "cleans" the pasture, which consists of cutting out the weeds, uprooting seedlings that the cows will not eat, and leaving commercially valuable seedlings to grow.

During the eight to twelve months preceding the cleaning, seed rain from nearby trees falls into the pasture grasses, and, in a humid climate, some of the seeds germinate spontaneously. The moment of truth for the seedlings arrives when a farmer or farm worker cleans the pasture after cattle have grazed it. She or he uproots or cuts the sprouting seedlings in an attempt to insure the continued dominance of forages for livestock in the field. If a seedling is of a commercially valuable species, the farm worker may leave it to grow. Over time, the selected seedlings grow larger, and a farmer-managed, partial regeneration of trees occurs in these fields. Ten to fifteen years later, the farmer may be able to sell the wood from the young trees to a local sawmill. Through this practice of selective culling in humid places, tree densities of anywhere from 90 to 350 stems per hectare accumulated over time in pastures.[11] As implied by this narrative, these carbon absorbing silvopastoral landscapes, as with Alexander Mather's agricultural adjustments, emerge during the thirty to forty years after the initial land clearing. They do so with particular vigor in humid settings such as the western Amazon basin.

This selective regrowth occurs in a nutrient poor environment. Soils in the tropics contain few nutrients. The standing biomass contains the largest pool of nutrients in these ecosystems, so when farmers clear the forests, it impoverishes the soils over time and slows pasture growth. The length of time for *gramalote* (*Axonopus scoparius*), the primary pasture grass in the Ecuadorian Amazon, to mature increased from eight to twelve months during the first thirty years after deforestation. The lengthening of the time to maturity for the pasture grasses reduced the frequency with which smallholders could rotate cows through a pasture. In so doing, the diminished fertility of the cleared lands, called the *Barbecho* crisis in the Bolivian Amazon, reduced the number of cattle that a smallholder could support on his pastures.[12] Household incomes declined proportionately.

Building wealth challenged the region's small-scale cattle ranchers in other ways as well. Without secure titles to land, many colonists and every Shuar could not obtain loans from banks to purchase cattle, so frequently they had no cattle to graze on recently cleared pasture lands. Under these circumstances, smallholders would rent their pastures to people with insufficient pasture for their cattle. Many new landowners planted *naranjilla* (*Solanum quitoense*), a citrus fruit, and did well for several years until pests discovered the new plantings and decimated the crops. Small-holders with cattle faced additional economic distress in the late 1990s when Ecuador replaced its national currency with the American dollar. To placate urban consumers buffeted by price rises after the introduction of the dollar, the national government allowed duty-free imports of inexpensive Argentine beef. The imports reduced the prices of Ecuadorian beef to the point where smallholders felt it necessary to liquidate their herds in order to pay off their loans from the banks and prevent foreclosures. These adverse economic events spurred outmigration from the region by the mestizos, beginning in the early 1990s with the failure of the *naranjilla* crops. While these economic crises did compromise the economic condition of cattle ranching households, many mestizo and Shuar households did acquire secure land titles during the last two decades of the twentieth century.

In this humid, economically turbulent context, smallholders had to make decisions about what to do with the tree seedlings that they encountered daily while cleaning the pastures after grazing by cattle. Seedlings from economically valuable species such as cedars (*Cedrela odorata*) and laurels (*Cordia alliodora*) promised another stream of income for farmers in fifteen to twenty years when wood from these trees could be sold to local sawmills. Under these circumstances, farm workers allowed economically valuable, spontaneously sprouted seedlings to continue to grow. Over time young trees from the selected species accumulated in pastures. Figure 6.1 depicts a silvopasture in the Ecuadorian Amazon with a dense but not atypical stand of trees.

The smallholders had created a silvopasture without putting in additional labor to plant the trees. The spontaneous germination of tree seedlings in the pastures depended on ample rainfall and patches of nearby rainforest to provide the seed rain. The shade from the new trees did not reduce the productivity of the pastures because the predominant pasture grass in the region, *gramalote*, was shade-tolerant. In effect, these spontaneous silvopastures generated a second stream of income for smallholders without requiring the labor usually associated with planting trees. Local sawmill owners added to the value of the trees in pastures when they

FIGURE 6.1 A typical silvopasture in the Ecuadorian Amazon, 2015. Photo by the author.

figured out how to make furniture out of tree species such as winchip (*Pollesta discolor*) that commonly sprouted in pastures.

Surveys of ninety-nine farms in four communities in a cattle-ranching region of Ecuadorian Amazon provided data on tree densities and the tendencies of different farmers to promote the creation of silvopastoral landscapes on their farms. Table 6.2 describes the variations in tree densities across pastures, broken down by communities. Several other factors added to the attraction of silvopastoral landscapes. Farmworkers would allow regrowth to occur along watercourses, which, by protecting water sources from the sun, would prolong their utility in the rare instances when ranchers in this region had to water their cattle. Ranchers thought that cattle benefited from the shade provided by the trees. Litter from the trees also added nutrients to the soils of the silvopastures. Appreciation of the advantages of silvopastoral landscapes concentrated among the younger generations in households. Farms with extensive silvopastoral stands of trees often had young persons who changed the cattle each day.[13]

The data on tree seedlings and rental arrangements suggests that the typical time horizon of land users mattered in the farmers' willingness to allow tree seedlings to grow. Rented pastures contained sprouting seedlings, but the renters, who typically cleaned the pastures, cut the seedlings out because they believed that, if the seedlings were allowed to grow, they would shade out the pasture, and reduce its productivity in the short run when the renter was using it. For this reason, pastures under the care of renters rarely ever became silvopastures. Economically disadvantaged groups such as the Shuar, who rented out their pastures because they did not have any cattle, had fewer trees in their pastures (see table 6.2).

Conversely, when landowners cleaned pastures, they anticipated living on the farm until the trees grew large enough to generate revenue for the household through their sale to sawmills. These projections assumed, of course, that the smallholder had secure land tenure. Because agrarian reform officials had made this zone the target of a new land settlement scheme during the last two decades of the twentieth century, most small-holders, Shuar as well as mestizo, did have secure titles to land (see table 6.2).

During the 1980s, after extensive clear cutting, the aboveground carbon pools for many pastures must have approached zero. Given this assumption and the above ground C readings of ~ 30 AGB/ha, in the communities with the most tree stems in 2011, estimates of the additional carbon

TABLE 6.2 Silvopastures: Patterns Among Smallholders in the Ecuadorian Amazon

Community	Sinai (M)	Huamboya (M)	Proano (M)	Santa Isabel (S)
Number of farms	23	24	23	30
Stem density (#/ha)	204	344	225	85
Palm density (#/ha)	18	37	2	5
Aboveground carbon (Mg/ha)	31.4	25.3	17.1	6.1
Renters (percentage of all households who rent out their land for grazing)	36.3	8.3	9.0	73.3
Average size of farms (ha)	43.5	42.9	20	14

Note: M = mestizo community, S = Shuar community

Sources: Adapted from Megan McGroddy et al., "Effects of Pasture Management on Carbon Stocks: A Study from Four Communities in Southwestern Ecuador," Biotropica 47, no. 4 (2015): 407–15; and Amy Lerner et al., "The Spontaneous Emergence of Silvo-pastoral Landscapes in the Ecuadorian Amazon: Patterns and Processes," Regional Environmental Change 15, no. 7 (2015): 1421–31.

sequestration in emerging silvopastures might have approached a ton of carbon per hectare per year. If the price for carbon in the offset markets remains about $15 per ton and the smallholders average twenty hectares of participating pasture in a REDD+ like program, the payments to an average farmer for the carbon absorbed by the sprouting trees would come to approximately $300 per annum. For Ecuadorian smallholders in this particular zone, REDD+ payments of this magnitude would earn them about 10 percent of their annual household income, a not insignificant sum of money. This conclusion about the economic feasibility of REDD+ programs for silvopastures would only pertain in the humid, tropical settings where biological communities sequester large amounts of carbon. The size of the REDD+ subsidy to silvopastoral ranchers would depend on how quickly landowners decided to harvest the trees in their pastures. Quick harvests of trees, after fifteen years, for example, would be less deserving of REDD+ subsidies than harvests after forty-five years of growth.

These calculations remain hypothetical because no environmental NGOs or government agencies demonstrated any interest in implementing a REDD+ program for silvopastoralists in the Ecuadorian Amazon. The lack of initiative seems particularly egregious given that many of the small farmers in this region have already begun sequestering carbon on their lands without any outside interventions. Carbon sequestering programs have not emerged in settings such as the Ecuadorian Amazon that would seem conducive to them. Their absence most likely stems from the lack of connective tissue between local land users and global environmentalists that corporatist processes typically provide.

A similar project in Colombia, described briefly in the introductory chapter, began under less promising circumstances than in Ecuador, but it made substantial gains during its first few years of operation. Trees did not regenerate naturally in the Colombian pastures, so the participating small and medium scale ranchers had to plant the trees themselves. Despite the additional labor requirements the smallholders did plant the trees, and the trees had the intended effect. Avian biodiversity increased significantly in the project area, and the growth of the trees sequestered, on average, 1.5 tons of carbon per hectare per year.[14] The Colombian project had substantial financial support from the Global Environmental Facility and the United

Kingdom. With these funds, Colombian officials convened workshops that mapped out the implementation of the program with farmers' representatives in two areas of southeastern Colombia. With the overseas funds, program managers established a payment for environmental services (PES) program that rewarded landowners for the increases in avian biodiversity and sequestered carbon. The numbers of participating landowners increased by 90 percent through the first six years of the program. The project seemed so successful that its sponsors decided to expand it. The Colombian government started successor programs in five additional locations close to protected areas in Colombia. The organizational partners of the project multiplied during this second stage, including environmental NGOs and the national association of cattle ranchers.

While the socioecological conditions did not differ much between the Ecuadorian and Colombian projects, the politics differed dramatically between the two places. These political differences seem attributable to the larger and more robust organizational coalition that implemented the project in Colombia. It contained a full range of local and global participants. The contrasting experiences in the two countries underscores the pivotal importance of corporatist political processes in implementing natural climate solutions.

Babassu Silvopastures in Brazil's Eastern Amazon Basin

The most common silvopasture in the eastern Amazon basin features babassu palms (*Attalea speciosa*) that have spread with the rapid expansion of pastures in the eastern Amazonian province of Maranhao during the second half of the twentieth century.[15] By the mid-1980s, secondary forests dominated by babassu palms covered 200,000 km^2 in the Amazon basin, and babassu-dominated silvopastures occupied another 200,000 hectares of land along the eastern edge of the Amazon basin.[16] Babassu spreads through the subterranean extension of roots from the tree. The belowground propagation of babassu protects the plant from fires set by shifting cultivators. As a result, babassu often dominated the secondary forests that reoccupied old sites of shifting cultivation where cultivators had used fires to clear land. With growth in rural populations and a corresponding expansion in shifting cultivation during the second half of the twentieth century, the extent of babassu dominated secondary forests grew in the Amazon basin. Figure 6.2 depicts a typical babassu agroforest in Maranhao, Brazil.

Another human driver, cattle ranchers, appeared during the 1960s. Between 1967 and 1984, the Brazilian state created a series of fiscal incentives that spurred dramatic increases in cattle ranching in the Amazon basin. A rapid, fire-assisted conversion of forests into babassu silvopastures occurred, especially in zones where ranchers planted invasive *Jaragua* grasses (*Hyparrhenia rufa*). The spread of the invasive accelerated the spread of babassu because *Jaragua*, when it is heavily grazed by cattle, does not suppress the emergence of babassu seedlings in pastures.

Smallholders welcomed the growth of babassu because the fruit of the tree had a wide range of uses that generated streams of income for the people who gathered or picked it. The hard shell of the fruit provides the raw material for charcoal. The kernel of the fruit yields a vegetable oil for cooking or for biofuel. The pulp of the kernel provides the basis for a feed

FIGURE 6.2 A Babassu agroforest. Photo by Michael Balick.

cake that livestock consume.[17] As babassu silvopastures expanded, it became customary each year for poor rural people, mostly women and children, to collect the fallen fruit from the fields and earn a livelihood from their share of the sale of the fruits to processors. By the late 1980s, 300,000 Brazilians earned a living from the gathering and sale of babassu fruit in eastern Amazonia.[18]

The simultaneous expansion of cattle ranching during the 1960s, 1970s, and 1980s concentrated landownership in the hands of large-scale cattle ranchers in the babassu zone. The large landowners, fearful of challenges to their landownership, focused their energies on maintaining exclusive control over their lands in the face of the large numbers of landless peasants who lived nearby.[19] To this end, the large landowners challenged the rights of the rural poor to collect babassu fruit from the landowners' fields. The landowners contracted with their own workers to gather the fruit from their fields. The rural poor responded to the landowners' exclusionary actions through collective action. Groups of collectors formed agro-extractive cooperatives and challenged the large cattle ranchers' right to prevent the rural poor from harvesting babassu fruit in the cattle ranchers' silvopastures.[20] In several places the collective action led to the creation of community-owned babassu silvopastures.

As a countermeasure, some large landowners sowed *Brachiaria* grasses in their pastures. The *Brachiaria* genus of grasses has high protein content, and, as an aggressive ground cover, it prevents babassu from regenerating in silvopastures. The conversion from *Jaragua* to *Brachiaria* pasture grasses by large landowners made it possible for them to increase the productivity of their cattle ranching operations at the same time that they excluded babassu fruit collectors from their lands. The loss of babassu trees from these converted tracts of pasture lessened the aboveground and belowground stocks of carbon in these lands.

During this same period, some smallholders with less than fifty hectares of land began to engage in small-scale cattle ranching and at the same time became active participants in the Jaragua-babassu silvopastoral trade. The engagement of smallholders in the babassu economy occurred despite insecure titles to land. Smaller-scale cattle ranchers with less income appreciated both the diversity and certainty of income streams from the

babassu palms. The relatively small size of the profits available to participants in the babassu trade may have diminished the enthusiasm of large-
scale cattle ranchers for participating in babassu commerce. These countervailing political dynamics between large and small landowners created
a compartmentalized babassu economy in Maranhao over time. The large-
scale cattle ranchers eliminated large swaths of silvopastures at the same
time that small-scale cattle ranchers, despite insecure titles, promoted the
continued existence and renovation of the babassu silvopastures.[21]

The history of recent land use changes in Maranhao has featured long
periods of environmental deterioration interspersed with episodes of
change that suggest the possibility of an ecological recovery. The shifting
cultivators and the rapid spread of cattle ranching enabled the emergence
of babassu-dominated silvopastures and the rise of an economy based on
the collection of babassu fruits in the pastures and secondary forests. The
countervailing actions of the large-scale cattle ranchers caused a surge in
greenhouse gas emissions that, over time, reduced the aboveground carbon
pool from a range of 141–571 Mg/ha in primary rainforests to a range of
2–16 Mg/ha in the *Brachiaria* pastures. These events tell a story of environmental deterioration over time.

The persistence of the babassu silvopastures in parts of Maranhao does
suggest a possible counternarrative of environmental recovery. Although
studies of carbon sequestration in the babassu silvopastures have not been
done, studies of similar landscapes make it possible to draw some conclusions about the gains in mitigation from the spread and persistence of the
babassu silvopastures. They have modest but still quite discernible mitigating effects. In the rare but not inconceivable conversion from a degraded
pasture without trees to a silvopasture, the aboveground carbon pool would
increase from a range of 2–16 Mg/ha to a range of 1–45 Mg/ha.[22] The owners of these lands would presumably qualify for REDD+ payments for the
carbon sequestered through the conversion of their lands to silvopastures. The combination of the monetary rewards from the babassu products, coupled with the REDD+ payments, would increase the incentives to
expand the extent of the babassu silvopastures. Given that the gains from
babassu commerce concentrate among landless peasants, this trajectory

of agricultural change would, in addition to sequestering some carbon, distribute income downward.

The historical experience with babassu and the class-based contention between poor babassu collectors and wealthy landowners suggest that class conflicts can obstruct the implementation of large-scale carbon sequestration efforts. The most straightforward resolution of the intertwined class and environmental issues would entail granting rights of access to the babassu stands through an agrarian reform that would award land to the babassu collectors and compensate the expropriated landowners. Through this type of policy, the owners of the babassu stands would become stewards of the environment. Stewardship that would promote forest expansion or restoration would seem most likely when people have long-term vested interests in land, as usually happens when they own the land that they work. The contentious history of agrarian reform in Latin America lowers the likelihood of these reforms, absent a widespread popular mobilization on their behalf.

Woodland and Parkland Expansion in Semiarid Sub-Saharan Africa

The spontaneous sprouting of seedlings in wet climates eliminates the labor of tree planting, so the spread of silvopastoral landscapes often occurs in humid places. In arid settings, farmers often have to plant the trees, so the labor costs of expanding silvopastoral landscapes would be higher and the density of trees would most likely be lower.[23] While the gains in carbon sequestration in these settings would be lower than in more humid environments, they would not be negligible. In addition, payments for carbon sequestration in arid landscapes, while small, would go to the poorest of the poor.

Silvopastures overlap broadly with dry forests and parklands in that they all contain grasses and relatively sparse stands of trees, often occur in semiarid environments, and cover immense areas of the tropics. Dry forests represent 42 percent of the world's tropical forests.[24] Parklands feature mature trees scattered in cultivated or recently fallowed fields, so they resemble sparse forests (see figure 6.3). Ninety percent of the agricultural parklands also serve as pastures for livestock,[25] so they are silvopastures as well. A belt of these silvopastoral lands stretches across Africa in the Sahel just to the south of the Sahara Desert. The Miombo woodlands south of the Congo River basin also feature large expanses of sparsely wooded landscapes that inhabitants farm, log, or burn in pursuit of livelihoods for themselves. Smallholders have converted 50 percent of the dry forests in East and West Africa for agricultural use.[26] These sparse forests cover extensive areas. In Mali, for example, parklands cover 90 percent of the agricultural lands in the country.[27]

During the middle decades of the twentieth century, immediately before and after the colonies achieved independence, the woodlands retreated in the face of droughts, rapid rural population growth, and property laws that discouraged tree planting. During the 1980s, these forest cover trends turned around, and the parklands began to expand in extent.

The predominant trees in the parklands have many uses. For example, the leguminous *Faidherbia albida* increases the productivity of cereal crops

FIGURE 6.3 Parkland, Sahel region. Photo courtesy of the UN Food and Agricultural Organization.

under the *F. albida* canopy by more than 100 percent compared with the yields from these cereal crops outside of the *F. albida* canopy. The disparity between cereal yields under the *F. albida* canopy and outside of it grows even larger during a drought. The yield increases around *F. albida* trees are also especially pronounced when the fertility of the soils is low. Other parkland trees typically provide a wide range of benefits to smallholder families. *Vitellaria paradoxa* yields nuts that, in processed form, become vegetable oil or butter. In addition, families use branches from *V. paradoxa* as fuelwood and, in processed form, as charcoal. The fruits from a wide range of trees in parklands enter diets occasionally. They benefit children in particular by diversifying the nutrients in their diets.[28]

The biogeographical environment of most parklands fluctuates in very consequential ways for farmers and villagers. Prolonged periods of drought, for example, between 1968 and 1973, led to crop failures, famine, and elevated tree mortality in the Sahel. High winds blow away recently sown seeds, so farmers have to reseed fields repeatedly. Soils in the Sahel tend to be relatively infertile, so adequate crop yields often require nutritional supplements in the form of litter from nearby trees and dung from livestock.[29] As noted earlier, the yields from cereal crops double when grown in the shade from the trees. Perhaps because shade reduces soil temperatures, it retards the loss of soil moisture after rainfall.[30] Farmers also dig planting pits to channel the runoff from rains toward recently planted seedlings. In some settings, farmers construct bunds to diminish runoff and direct it toward croplands. While some trees regenerate naturally, farmers plant a large fraction of the trees nearest to their homes. Considered together, these practices suggest that additional labor by smallholders has proved crucial in fostering the regeneration that has occurred in semiarid woodlands and parklands.

Rapid population growth in rural areas has created plentiful supplies of labor. The annual increment in population in Niger in 2008, for example, reached 3.7 percent.[31] Under these circumstances, population density in rural areas correlated positively with tree densities. The proportion of planted trees increased with proximity to homes. At the same time, the pressure of wood collectors on fuelwood supplies increased. During the 1970s and 1980s, fuelwood collectors and merchants stripped periurban

144 • Agroforests I

areas of wood in order to meet the growing demands of urban residents for
wood and charcoal. Denuded areas, plainly visible on satellite images,
grew up around some cities. Close to cities, some smallholders have begun
to engage in contract tree planting. Urban investors provide fertilizers to
nourish planted trees. Years later, the smallholder sells the trees, at a dis-
counted price, to the same investors who then use the timber as construction poles or as raw material for charcoal.[32]

Fundamental revisions in "tree tenure," the laws and customs that determine the ownership of trees, spurred the turnaround in forest cover trends
in the Sahel during the 1980s. During the first half of the twentieth century, foresters with the colonial regimes had applied the forestry laws of
the occupying powers to the colonies. In French colonies such as Niger,
trees became, as in France, property of the state. This pattern of tenure challenged the smallholders' practice of claiming ownership of trees that they
had planted and using forest products from the trees that they had planted.
In doing so, the colonial-era laws discouraged smallholders from planting
trees.

The colonial rules remained in place during the first few decades after
independence. Beginning in the 1980s when the debt crises reduced central government activities, local councils of village elders in countries such
as Niger began to exert more control over tree tenure. NGOs began to promote farmer managed natural regeneration in which committees of village
elders oversaw the implementation of local agricultural improvements such
as tree planting.[33] At the same time, smallholders had begun to follow the
norm that the person who plants the tree controls the harvests from the
tree. In other words, "labor creates rights."[34] In 1985, the central government in Niger reformed its tree tenure laws so that they conformed to the
locally held norm that whoever plants the tree owns it. This shift in laws
reinforced the tendency for smallholders to expect tangible rewards from
planting a tree.[35] Farmer-managed regeneration of trees had begun to shape
processes of landscape change in these semiarid, silvopastoral districts.

In the dry Miombo woodlands to the south and east of the Congo basin,
political controls over the fuelwood and charcoal trade evolved during the
1980s in ways similar to the political dynamics that characterized controls
over the parklands in the Sahel. Village councils in Tanzania established

controls over the granting of permits to extract fuelwood from the woodland commons that surrounded their villages.[36] By limiting the number of permits, the councils imposed controls over the extraction of fuelwood from community forests. These restrictions made the fuelwood trade more sustainable.

Similarly, village oversight committees in the Sahel acquired the authority to review farmer managed regeneration in the parklands. Under the local authorities, the numbers of trees in parkland districts of the Sahel increased substantially, from a range of three to seven trees per hectare in 1980 to a range of thirty to forty trees per hectare in 2000.[37] In Niger, villagers rehabilitated five million hectares and cereal yields increased by one hundred kilograms per hectare over the same period. The trees themselves grew larger in the rehabilitated areas than they had before reforestation. Even so, the additions in aboveground carbon were modest, with the aboveground biomass totaling only 4.5 tons per hectare in some of the rehabilitated parklands. Still, the numbers of new trees become considerable given the large areal extent of the regenerated parklands. If the increments in trees average out to 40 trees per hectare over 5,000,000 hectares in eastern Niger, the number of trees increases by 200,000,000. Summed across other states such as Burkina Faso with similar programs, the amount of sequestered carbon in these regenerated parklands becomes significant. It would be easy, however, to overstate these effects. The semiarid environment slows plant and carbon growth. Regrowing Miombo woodlands sequester only.7 tons of carbon per year per hectare, substantially less than the 1 ton of carbon sequestered per year per hectare in the humid silvo-pastures of Ecuador.[38]

The growth in the numbers of trees increased the forage available for livestock. In some places, the increased vegetation has led to sedentarization among herders, which in turn has contributed to increased cooperation between farmers and herders.[39] The herders and farmers agreed on plans for the distribution of dung on fields in the parklands. Regenerating parklands also addressed some equity issues involving women. They had long had responsibility for the collection of firewood. Increases in tree density, especially close to compounds, reduced the amount of time it took to collect firewood. In Niger, the amount of time necessary to collect the firewood for the household declined from three hours to thirty minutes during the 1990s.[40]

Regenerating forests begin to yield dividends for smallholders as soon as two to three years after planting, when some saplings grew large enough that they could be pruned for use as firewood. Other transformations, such as the increased soil fertility at the base of *F. albida* trees, can take twenty to thirty years to appear.[41] The changes in governance at the village level in Niger that spurred the turnaround in forest cover changes also unfolded over several decades. This history implies that the time frame for incentivizing farmers to sequester carbon in dry forests could be ten years or more.

The mutually supportive actions among politicians, farmers, and nongovernmental organization (NGO) activists in Niger during the 1980s and 1990s resembles a corporatist pattern of political activity, and it did achieve visible gains in forest cover. The success of the reforestation initiatives in Niger in the aftermath of the severe droughts in the 1960s and 1970s spurred a more wide-ranging reforestation program after the millennium. In 2007 eleven African nations ranging from Djibouti in the east to Senegal in the west agreed to rehabilitate forests and scrub growth in the Great Green Wall (GGW) initiative. A belt of planted trees a hundred million hectares in extent would stretch from the Red Sea to the Atlantic Ocean. The GGW forest would forestall the advance of the Sahara Desert southward. In 2014 the Food and Agriculture Organization initiated an "Action Against Desertification" program to implement the GGW with support from the Turkish government.

The initial accomplishments of the GGW initiative have been disappointing. Only 4 percent of the project's original goal for restored forests has been achieved during the first fourteen years of the project. Political conflict has slowed or stopped the restoration effort in one half of the project area.[42] The associated violence has intimidated participants and splintered the coalitions, like the one in Niger, that had fueled earlier restoration efforts. Farmers left the danger zones, and NGO personnel refused to travel to insecure areas. The collapse of the coalitions and resulting absence of restoration underscores in a negative way the importance of the coalitions to the restoration effort.

Silvopastures in Comparative Historical Perspective

Table 6.3 summarizes the diverse socioecological conditions that characterized the three case studies of silvopastoral expansion described here. Of the three cases, the events in Niger most closely approximate the corporatist pattern of diverse groups coming together to make reforestation happen. Each of these cases demonstrates in different ways the salience of political processes in restoration efforts. The dramatic differences between Ecuador and Colombia in external assistance and project performance illustrates the pivotal influence that external assistance can play in getting reforestation processes going. In the Brazilian case, class conflict prevented the spread of silvopastures and the increments in environmental services that would have accompanied the spreading silvopastures. In the Sahel, the assembly of a corporatist coalition during the last two decades of the

TABLE 6.3 Silvopastures and Woodlands in Africa and Latin America

	Upper Amazon	Lower Amazon—Babassu Palms	West African Parklands
Rainfall	2,000–4,000 mm/yr	~1,500 mm/yr seasonal rains	400–900 mm/yr seasonal rains
Tree density/ha	85–344	~98	20–50
Means of spread	Seed rain	Subterranean root system	Planted trees and seed rain
Biodiversity	Medium	Low	High
Food security	Secure	Insecure	Insecure
Land distribution	Egalitarian	Concentrated among the wealthy	Egalitarian
Land tenure	Secure	Insecure	Insecure
Wealth	Lower middle class	Concentrated; otherwise impoverished	Impoverished
Fuelwood?	No	No	Yes
Political coalitions?	No	No	Yes
Carbon sequestration?	Some	Little	Some

Sources: Amy Lerner et al., "The Spontaneous Emergence of Silvo-pastoral Landscapes in the Ecuadorian Amazon: Patterns and Processes," *Regional Environmental Change* 15, no. 7 (2015): 1421–31; Anthony Anderson, Peter May, and Michael Balick, *The Subsidy from Nature: Palm Forests, Peasantry, and Development on an Amazon Frontier* (New York: Columbia University Press, 1991); Chris Reij, Gray Tappan, and Melinda Smale, "Agroenvironmental Transformation in the Sahel: Another Kind of 'Green Revolution' " (Washington, DC: International Food Policy Research Institute).

twentieth century generated growth in forest cover in an inhospitable environment. Increases in ethnic and religious conflict after 2000 disrupted the coalition and diminished restoration efforts.

The data gathered by Robert Zomer and his associates document the recent expansion of silvopastures and agroforestry. The case studies in this chapter identify the political processes that have encouraged the spread of silvopastures in some instances and blocked their spread in other instances. Augmented outside assistance can build on local processes such as those in Ecuador, making it more likely that these local processes will expand in scale.[43] In addition, in the case of the African parklands, external assistance would have an unmistakable redistributive effect, contributing to sustainable development with equity.

7

• • • •

Agroforests II

RESTORING AGROFORESTS IN THE HUMID TROPICS

Cultivators practice agroforestry when they interplant commercially valuable woody perennials such as coffee with other trees and bushes that enhance the yields from the valuable perennials. Vegetation in an agroforest typically has multiple strata, with a cultivator planting the understory or the overstory of a forest with a valuable, fruit bearing bush or tree. Vegetation in an agroforest takes on a tiered appearance, as in figure 7.1, with different fruit bearing plants at different strata in the garden or forest. Coffee and cocoa cultivators create agroforests by planting the understory of an old growth forest with the fruit bearing tree or bush. Smallholders who practice other kinds of agroforestry manage their plantings and extraction to favor particularly useful trees in the overstory such latex-producing rubber trees, fruit-bearing acai palms, or resin-producing damar trees.[1] By integrating production from trees with production from bushes and crops, agroforestry diversifies and sustains overall yields from the land.[2]

Because the focal plants in an agroforest typically do not provide fruits to farmers for the first three to six years after planting, agroforestry represents a long-term investment. Managing an agroforest requires a long-term perspective, with a view of the farm as an enterprise sustained by the farmer and the surrounding plant community for thirty to forty years. Mature agroforests deliver streams of income to smallholders and sequester substantial amounts of carbon. These attributes add value to landscapes

FIGURE 7.1 Multitiered garden, Ecuadorian Amazon. Photo by the author.

during an era of climate change, and they raise pressing questions about the circumstances that promote the emergence of poverty-alleviating, carbon-sequestering agroforests. This chapter tries to answer these questions.

Early in the twenty-first century, more than 1.2 billion people worldwide cultivated more than a billion hectares in different types of agroforestry.[3] Tree crops occupied about 10 percent of all arable land worldwide.[4] Almost all of the farmers who practiced agroforestry were smallholders, working less than four hectares of land. In the large, densely populated rural areas of South Asia, East Asia, and sub-Saharan Africa, an economic logic tied smallholders to agroforestry. By making it possible to harvest a more diverse set of crops, agroforestry reduced the risks faced by smallholders who produced crops such as rubber, cocoa, and coffee that sold in volatile world markets.

Agroforestry almost always has a multitiered structure, with most of the economic returns coming from a plant in the understory or the overstory. On occasion, as in Brazil, one plant such as rubber yields a return in the overstory while another plant, cocoa, produces a return from the understory.[5] Agroforestry also features crop rotations such as one that begins in the initial stages of cultivation with annual row crops such as rice mixed in with longer-lasting plants such as rubber trees. When the rubber trees grow larger, smallholders stop cultivating rice on these lands and begin to tap latex from the trees. During this later phase, the composition of harvested products shifts again. Smallholders begin to harvest products such as aromatic pandanus leaves (*Pandanus amaryllifolius*) from the rubber forest. Eventually, the old rubber trees produce less latex, and smallholders allow the land to go fallow until they began a new cycle of cultivation with newly planted rubber trees.[6] Some smallholders have embarked on similar cycles of cultivation with other commercially valuable trees such as oil palm.

The multistrata vegetation of an agroforest represents a middle way between environmental service–rich secondary forests and environmentally degraded croplands. This middle path has variable environmental impacts, as illustrated in table 7.1. An expansion in agroforestry provides significant increments in environmental services only to people and places where it replaces degraded, short-fallow agriculture. Our case studies will focus on three tree crops, rubber (*Hevea brasiliensis*), cocoa (*Theobroma Cacao*), and coffee (*Coffea*). The last case study in the chapter looks at domestic forests, a particular type of species-rich agroforestry that occurs in close proximity to smallholders' homes.[7]

Shifting Cultivation, Jungle Rubber, and the Spread of Plantations

At the beginning of the twentieth century, most Indigenous groups in insular Southeast Asia practiced shifting cultivation in which they planted upland rice, tubers, and fruit trees in small, recently cleared fields. After

TABLE 7.1 Agroforest Transitions and Their Carbon Consequences

	Net carbon sequestration	Place	Source
Short-fallow swidden to jungle rubber	Yes	Sumatra	Pye-Smith 2011
Long-fallow swidden to jungle rubber	No	Sumatra	Pye-Smith 2011
Jungle rubber to rubber/ oil palm plantations	No	Southeast Asia	Ziegler, Fox, and Xu 2009
Primary rainforest to cocoa agroforest	No	Côte d'Ivoire	Ruf 2001
Grasslands to cocoa agroforest	Yes	Sulawesi	Ruf 2001
Sun-filled to shade-filled coffee	Yes	Colombia	Rueda and Lambin 2013
Primary rainforest to shade-filled coffee	No	Honduras, Vietnam	Nagendra, Southworth, and Tucker 2013
Kitchen garden to domestic forest	Yes	Kenya	Tiffen, Mortimore, and Gichuki 1993

two or three years of sowing more rice and harvesting fruits and tubers, they began to work another plot of land, so in this sense cultivation shifted periodically to new lands. After Indigenous people abandoned a field, they let it lie fallow for fifteen to twenty years, a practice that cultivators have called a long fallow. Given a humid climate, vigorous regrowth occurred in the long-fallowed fields. Over time, the decaying and reemerging vegetation in long fallows restored the fertility of the soils and prepared the ground for another cycle of cultivation. With population increases in rural areas of the Global South during the post–World War II era, long fallowed fields became harder to find and shorter fallows of seven to ten years became more common on the lands worked by shifting cultivators.

The arrival of rubber in Southeast Asia at the beginning of the twentieth century disrupted these cycles of shifting cultivation. Shifting cultivators planted rubber trees that they would tap for latex in groves of two to three hectares. The sale of the latex provided sedentarizing shifting cultivators with an income. By the beginning of the twenty-first century, 90 percent of the world's rubber came from Asia, and 90 percent of that production came from smallholdings, often under two hectares in extent.[8]

Smallholders in Southeast Asia harvested rubber from more than ten million hectares in 2000.[9] To establish a rubber agroforest, cultivators interplanted rubber tree seedlings with upland rice. The first crops in these fields, along with the rice and the rubber seedlings, had an aggressive, pioneering quality. Crops such as bananas and papaya grew quickly and created shady and humid microclimates in which tree species such as rubber, fruit trees, and palms germinated easily.[10] During this post-pioneer phase, coffee, cloves, and peppers also grew in abundance in the understory. Fifteen to twenty years after the establishment of the garden, trees in the overstory, such as rubber, formed a high and closed canopy. The wide range of useful species made for a complex agroforest. The smallholders created paths from rubber tree to rubber tree which they followed when they tapped latex from the trees. The multiple strata in these agroforests, each with a commercially useful product, mimicked the structure of a rainforest and persuaded observers to refer to these rubber gardens as "jungle rubber."[11]

A survey of rubber gardeners in Sumatra in 1998 and 1999 captured the essential features of rubber agroforestry. The gardeners had landholdings that averaged 6.4 hectares, of which only 2.2 hectares contained rubber trees. The smallholdings contained an average of 525 rubber trees per hectare, so jungle rubber operations often had more than a thousand rubber trees. The smallholders used very few external inputs such as fertilizer to augment yields of latex.[12] About half of the smallholders relied on family labor to care for and tap the rubber trees. One-half of the smallholders reported labor scarcities. Seventy per cent of their household income derived from the sale of latex from the trees.

The rubber gardens provided a crucial element in the livelihoods of smallholders in insular Southeast Asia. The smallholders would tap the trees when the needs of their households required cash purchases, so in this respect rubber tapping became an essential feature in the subsistence economy of the cultivators' households. Smallholders maintained a relatively constant level of production from their agroforests from year to year through *sisipan*, a practice in which they replaced old, declining rubber trees, one by one, with rubber tree seedlings. Smallholders harvested the bulk of the rubber in Indonesia during the late twentieth century. During

FIGURE 7.2 A young coffee plant in an understory, Mindo, Ecuador. Photo by the author.

the 2000 to 2005 period, they controlled 85 percent of the rubber-producing land in Indonesia and produced 81 percent of the latex in Indonesia.[13]

While jungle rubber clearly produced higher and more stable incomes than shifting cultivation, further technological developments induced changes in rubber production that degraded the environmental services associated with it. During the 1990s, development banks and plantations began to promote monoclonal varieties of rubber, grown in plantation monocultures, that provided much higher yields than jungle rubber. The conversion to monoclonal varieties typically cost smallholders considerable amounts of capital because to establish the monoclonal plantations, growers had to clear stands of old rubber trees and replant the land with monoclonal seedlings. The labor associated with the clear cutting cost smallholders. The clear cutting also exacted a high toll in lost environmental services, including losses in sequestered carbon. Growers also did not want to cultivate the monoclonal varieties in association with other commercially valuable plants, so the monoclonal rubber fields contained very little biodiversity.[14] Given the higher yields of plantation rubber, it spread while jungle rubber contracted during the last decades of the twentieth century. Table 7.2 charts these changes in rubber cultivation in the Bungo District, a rubber-producing region in Sumatra, during the last few decades of the twentieth century.

These differences in the appearance of fields have had predictable and unfortunate consequences for the amounts of carbon sequestered in

TABLE 7.2 Jungle Rubber and Monoclonal Plantation Rubber: Trends Over Time

	1973	2005
Jungle (agro-forest) rubber	15%	11%
Plantation rubber	2%	27%

Source: Adapted from Charlies Pye-Smith, "Rich Rewards for Rubber? Research in Indonesia is Exploring How Smallholders Can Increase Rubber Production, Retain Biodiversity and Provide Additional Environmental Benefits" (Nairobi: World Agroforestry Centre, 2011).

rubber landscapes. As noted in table 7.1, the conversion from long fallow shifting cultivation to jungle rubber does not cause large changes in the amounts of carbon sequestered above ground. Both practices routinely sequester around 90 Mg per hectare of carbon in aboveground biomass.[15] Rotational rubber plantations, using monoclonal varieties with periodic clear cutting and replantings, retained an aboveground average of only 50 Mg per hectare of carbon over the years. Given these figures, a jungle rubber to rubber plantation conversion, as occurred in the Bungo District (table 7.2), and in many other places in Southeast Asia in the twenty-first century, has probably resulted in substantial greenhouse gas emissions.[16]

The scale of the land use conversion accentuated the surge in emissions. By 2009, smallholders and planters had converted an estimated 500,000 hectares from shifting cultivation to rubber plantations in the uplands of mainland Southeast Asia.[17] Greenhouse gas emissions of a similar magnitude have also occurred with the conversion of rubber agroforests into oil palm plantations.[18] The conversion to oil palm, like the conversion to monoclonal rubber, promises higher incomes for smallholders. Poorer smallholders persisted in some instances with jungle rubber in the face of the spreading rubber and oil palm plantations. These smallholders would plant small numbers of oil palms in plots of land that also contained rubber trees. While the small numbers of oil palms on these plots limited the smallholders' income, the diversity in products from the land cushioned the economic shocks suffered by smallholders from fluctuations in commodity prices. The diversified plant communities in these hybrid oil palm and rubber smallholdings also sequestered more carbon than the surrounding monocrop oil palm and rubber plantations.[19]

In sum, globalization, the growth in rural populations, and the extension of rural road networks in the second half of the twentieth century has led to a mix of environmental changes in rubber-dominated agricultural districts. In one sequence of changes, shifting cultivators, faced with growing scarcities of land, intensified agriculture. They extended the period that they cultivated plots of land and shortened the length of time that they let the land lie fallow after cultivation. Predictably, the shortened fallows reduced both the biodiversity and the biomass in these lands. Under these circumstances, smallholders who converted their lands from short fallow shifting cultivation regimens to jungle rubber probably increased both the biodiversity and the carbon sequestered in their lands (see table 7.1). In another, more recent sequence of changes, rubber agroforesters converted their jungle rubber operations into monoclonal rubber plantations or oil palm plantations. The conversions from long to short fallow shifting cultivation and from jungle to monoclonal rubber plantations both reduced the carbon stored in landscapes.

How could societies stem this drift among rubber agroforesters toward more intensive, plantation-like land uses that sequester relatively little carbon? Three activities come to mind. First, agronomists at the World Agroforestry Center have introduced new, improved practices for jungle rubber, with the idea that these practices will increase the returns to labor from jungle rubber and in doing so reduce the economic appeal of monoclonal rubber and oil palm plantations.

Second, ecocertification schemes could increase the appeal of rubber grown in biodiverse, jungle settings. In this instance, because jungle rubber smallholders engaged in sustainable practices, they would receive permission to sell the latex from their smallholdings in specialized markets at higher, privileged prices.[20] The recently created Global Platform for Sustainable Natural Rubber, while at present no more than an advocacy group, has an agenda that could transform it into an ecocertification institution for rubber producers.

Third, NGO activists could adapt REDD+ programs to jungle rubber practices. The increase in sequestered carbon when smallholders convert from short cycle shifting cultivation to jungle rubber would qualify them for REDD+ payments if they have secure titles to their land. Stemming the tide of conversions from jungle rubber to monoclonal rubber and oil palm plantations would seem to require the adoption of REDD+ like institutional arrangements. New paths to reforestation become feasible, but only after further deterioration in the environment makes the gains from a partial forest restoration through REDD+ enhanced jungle rubber more appealing to more people in more places.

Deforestation, Shade-Grown Cacao, and
State-Driven Rehabilitation Efforts

Global demand for cacao, like demand for rubber and oil palm, increased tremendously during the twentieth century. Worldwide cocoa production increased from 100,000 tons at the beginning of the twentieth century to 3.2 million tons in 2003–2004.[21] The increase in the volume of production spurred a comparable increase in the extent of cultivation. In just ten years, between 2000 and 2010, the area in West Africa under cultivation for cacao increased from 2,000,000 hectares to 3,000,000 hectares.[22]

Eighty percent of the world's cacao production comes from countries that border on the Gulf of Guinea, from the Congo in the east to Sierra Leone in the west. Intact, old growth forests covered the entire region before cocoa cultivation began during the second half of the twentieth century, so almost all of the subsequent expansion in cocoa agroforests came at the expense of old growth rainforests.[23] After 2000, the cocoa growing belt of West Africa began to experience some of the most rapid rates of deforestation in the tropics.[24]

Smallholders with two or three hectares of land have driven most of the expansion in cocoa cultivation in the humid zones of West and Central Africa. Young growers would move into a largely forested area, clear cut several acres of forest, and plant it with a new, more productive hybrid variety of cocoa. Although cocoa cultivation had traditionally taken place under a canopy of commercially valuable trees, younger growers after the turn of the century showed an increasing preference for growing cocoa in fields fully exposed to the sun. The young growers fashioned a strategy of land occupation that took advantage of a series of short-lived improvements in the productivity of the newly planted cacao. Adopting one of the new hybrid varieties of cacao increased the productivity of fields by 30 percent to 40 percent. Clearing old growth forests for cacao fields enabled growers to exploit the accumulated fertility of the soils beneath the forests, what economists have called "forest rent." Growing cacao in the full sun increased the productivity of the fields by an additional 25 to 30 percent. In addition, the flurry of activities associated with establishing the new fields strengthened growers' claims to the land in which they had invested so much of their labor. Furthermore, this agricultural expansion strategy did not require, at least initially, the use of expensive agricultural inputs such as herbicides or pesticides. In sum, these strategies maximized short-run returns for the growers.[25]

Norms about cultivating cocoa agroforests varied among different groups of smallholders. Natives of Côte d'Ivoire, long resident in the forested zones, were more likely to convert fallowed fields into cocoa agroforests.[26] In contrast, migrants to Côte d'Ivoire from Burkina Faso were much more likely to chop down old growth forests to establish a cocoa agroforest. Similarly, migrants from outside the cocoa-growing zone were more likely to purchase and try to rehabilitate abandoned, invasive infested grasslands. The migrants had higher discount rates than longtime residents. Typically, the migrants wanted to make an economic killing and return to their home villages in the near future, so they often opted for aggressive strategies of land use. In some instances, this stance made the migrants more likely to destroy old growth forests even though it required more work from them than clearing secondary forests. The migrants did so in order to capitalize on the accumulated fertility in the soils beneath the old growth forests. In another kind of aggressive strategy, migrants attempted to create cocoa agroforests on parcels of degraded land that longtime local residents would refuse to rehabilitate.

Many of the strategies of both migrants and longtime residents presumed the continued availability of abundant, open-access, long-forested land for new farms. By the onset of the twenty-first century, this assumption no longer seemed tenable. Unworked, unclaimed forest land had become scarce in the cocoa-producing belts of land. Economic and environmental deterioration in older, cocoa-producing regions reinforced the urgency of the search for a more sustainable strategy for cocoa production. The mortality of cocoa plants exposed to the full sun had increased significantly as the plants had aged. The increased mortality of cacao stemmed in large part from infestations of pests that the open sun environment of the new cacao fields encouraged. Finally, the aging of the cacao plants across the cocoa-producing belt reduced the productivity of the plants.

Forest regeneration has occurred in the old, degraded lands of the cocoa belts under several identifiable circumstances. First, farmers chose in some instances to invest in degraded lands where land users had few, if any, other economic opportunities. Degraded former cocoa plantations in Sulawesi, now overrun by an invasive grass, *Imperata cylindrica*, exemplify these places where investments in intensified agroforestry increased both the economic returns and the environmental services from land.[27] Farmers, for example, used herbicides to eradicate the invasive grasses. These investments paid off for smallholders if they had secure land tenure and if they stayed on the land long enough to benefit from the growth in production that occurred when the new trees begin to yield harvests.

Second, a modified version of the older, shade-filled agricultural regimen began to seem more reasonable to some growers. Shade-grown cocoa plants maintained their productivity for significantly longer periods of time than did the sun-grown plants. Pest infestations in the shade grown fields did not seem as intense. A meta-analysis of fifty-two studies that compared shade-grown with sun-filled cocoa systems did establish that the shade from the trees in the canopy reduced the productivity of cocoa plants by 25 percent. Nevertheless, the overall return from shade-filled cultivation exceeded sun-filled cultivation by a significant amount when the economic return included the proceeds from the sale of products from the other trees in the canopy of the cocoa agroforests. In addition, the food crops from the cocoa agroforests improved the food security of cocoa smallholders.[28]

Efforts at rehabilitating lands for cocoa agroforests have become more likely when smallholders can extract additional streams of revenue from the agroforests. Researchers have noticed several possibilities. In the old cocoa-producing zones of coastal Brazil, smallholders have begun to plant rubber trees in cocoa agroforests. The rubber trees become part of the tree canopy that shades the cocoa plants. In addition, the rubber trees provide an additional stream of income from the latex that the smallholders tap from the trees. Fruit trees play a similar role in cocoa agroforests near cities in Cameroon. Proximity to cities provides the owners of the cocoa agroforests with markets for the fruit that they harvest from trees in the agroforest canopy. For this reason, cocoa agroforests close to cities contain more fruit trees that do the agroforests located farther from cities.[29]

The superior overall economic performance of the shade-filled systems most likely explains the predominance of shade-filled cultivation among cocoa-producing smallholders in the Gulf of Guinea. In a producer survey across four countries (Cameroon, Nigeria, Ghana, and Côte d'Ivoire) 79 percent of all cocoa was produced in shade-grown systems during the late 1990s.[30] Cacao tolerates shade better than other tropical crops such as rubber, oil palm, and robusta coffee, which in turn makes cultivators more comfortable with shade-grown regimens for cultivating cocoa.

The superior environmental services associated with shade-filled cultivation suggests a rare win-win outcome in which the most economically advantageous pattern of cultivation also delivers more environmental services. The larger biomass present in the cocoa agroforests increased the volume of carbon sequestered in shade-grown systems. They sequestered 2.5 times as much carbon as did the sun-filled systems.[31] The shade-grown cocoa landscapes also maintained higher levels of biodiversity than did the sun-filled cocoa landscapes.

These findings suggest a return to shade-filled regimens of cultivation, but to date few growers have made the switch. Under these straightened circumstances, several courses of action have emerged. The growth of urban economies has increased the demand for wood for construction at the same time that exhausted lands have increased in extent in the cocoa belt. Smallholders have responded to the increased urban demand for wood by creating small forest plantations, one to two hectares in extent, on the old cocoa lands.[32] Tree planting brought its own complications with it. In Ghana loggers retained the right to harvest trees wherever they grew. These practices initially discouraged smallholders from establishing woodlots on their lands, but smallholders have recently figured out ways of evading these regulations, so tree planting on smallholdings has resumed.[33] REDD+ programs could, at the very least, compensate these smallholders for the extra carbon sequestered, provided that landowners delay the harvests from these woodlots.

Extension services have also begun to experiment with the use of fertilizers in cocoa groves. Persistent use of fertilizers did increase the productivity of the cocoa plants. States have also intervened with planting programs for new hybrid varieties of cacao in shaded fields. The governments of Côte d'Ivoire and Ghana both embarked on extensive replanting campaigns after 2011 that replaced old, unproductive cocoa plants with more productive hybrid seedlings.[34]

Agronomists have explored the likelihood, presuming adequate rainfall, of establishing shade-grown cocoa on exhausted agricultural lands. Efforts to rehabilitate "tired" cocoa lands have usually occurred in areas containing mosaics of land use, only some of which cultivators had rehabilitated for shade-grown cocoa. While the likelihood of widespread transitions from sun-filled to shade-grown cocoa seems low given the long time necessary to establish a canopy of trees for the shade grown cocoa plants, the planting of fast growing exotics might accelerate the formation of a canopy for shade-grown cocoa regimens. In some instances, cultivators will continue to exploit sun-filled fields until the growth of weeds renders these fields useless. In the outer islands of Indonesia, these degraded lands frequently contained heavy infestations of invasive grasses such as *Imperata cylindrica* that must be controlled if new plantings of food crops, tree seedlings, and cocoa plants are to take hold in an old field. The labor requirements for eliminating the invasive plants would seemingly have doomed these efforts before the widespread introduction of herbicides in the tropics.

The introduction of herbicides in the 1980s in Sulawesi reduced the labor required to eliminate the invasive grasses and to prepare the ground for replanting with cocoa and overstory trees.[35] The replanting does represent reforestation, and it sequesters carbon. It becomes advantageous to smallholders to replant and reforest these old fields when alternative sites for cocoa cultivation disappear. Some prospective sites with old growth forests become protected areas. Other potential sites become oil palm plantations. Still other sites disappear amid the roar of the loggers' chain saws and become, over time, degraded croplands. Under these straightened circumstances, the admittedly expensive rehabilitation of invasive infested grasslands becomes more palatable to prospective cocoa cultivators.

These rehabilitation efforts have varied across regions with the types of invasives that have infested degraded lands. In West Africa *Chromolaena odorata*, an invasive shrub from the Americas, has taken over abandoned cocoa lands. Migrants to Côte d'Ivoire bought up tracts of these degraded lands and developed routines for pulling out *C. odorata* that eradicated it from sites where they then went on to replant cocoa. The likelihood that smallholders would undertake rehabilitation efforts depended on the availability of unclaimed and unprotected old growth forests where they might cultivate cocoa. Where prospective cultivators could not convert old growth sites, they resorted to creating cocoa agroforests on degraded sites. In these instances, restoration efforts enhanced the biodiversity of these reclaimed landscapes and increased the carbon sequestered in them.[36]

Contextual factors frequently played a role in changing cocoa landscapes. Smallholders in Brazil found that planting cacao under a thinned canopy of an old growth forest cost them less than planting cacao on clear-cut lands. The shade reduced the productivity of cocoa bushes, but the lower costs of clearing and planting the thinned forest, as opposed to clear cutting it and replanting it with cacao, compensated for the foregone profits. When an economic crisis hit the cocoa industry in the late 1990s and farm workers left for the cities, the less labor-demanding shaded methods of cocoa cultivation made it possible to continue cocoa cultivation despite adverse price and labor trends.[37] Not so parenthetically, these more extensive methods left more trees standing and more carbon sequestered.

International corporate interventions have, as a goal, tried to spur the innovations outlined above. Indonesia, the world's fourth largest producer of cacao, launched a Sustainable Cocoa Production Program (SCPP) in 2012 with the participation of major international agribusinesses such as Mars, Nestle, and Cargill. The organizational profile of the SCPP resembles the organization of the oil palm industry. Smallholders receive planting materials and technical assistance from the agri-businesses. In return, smallholders sell their cacao to company-controlled Cocoa Development Centers that process the fruit. The intent of the SCPP was to increase the incomes of the smallholders and, through reforms of cultivation routines, reduce greenhouse gas emissions by 30 percent.[38]

These ecocertification initiatives began only after the devastating initial ecological impact on both intact and degraded rainforests of the wave of cocoa expansion that occurred during the late twentieth and early twenty-first centuries. Because cocoa expansion in West Africa, Central Africa, and Indonesia usually began with the destruction of old growth forests, it wiped out a wide range and large volume of environmental services (see table 7.1). With the disappearance of old growth forests and their fertile soils, smallholders have persisted in cultivating aging, shade filled groves of cocoa even as their productivity has declined. As a result, the poverty of the growers has persisted.

None of the above-mentioned ecocertification initiatives would involve additional deforestation, so their implementation would, in some instances, spur modest increases in carbon sequestration in cocoa-producing landscapes. They presume a strengthened state, with ties to local as well as international elites, that would funnel funds to smallholders to compensate them for their participation in REDD+ programs that sequester carbon in rehabilitated cocoa landscapes. In their organizational outlines, these activist states, with conduits for funds between elites and smallholders within agricultural sectors, would take a societal corporatist form.

Coffee, Other Tropical Commodities, and Waves of Ecocertification

Coffee and cocoa cultivation have followed similar historical paths in recent years. Interest in sun-filled coffee cultivation surged toward the end of the twentieth century followed by a renewed interest in the venerable, shade-filled techniques of cultivation. In most of Colombia and Brazil, coffee cultivation currently occurs in sun-filled monocultures.[39] The conversion of primary or secondary forests into sun-filled coffee fields in these places entailed large losses of above-ground and belowground carbon. In other places, such as Honduras during the 1990s, the requirements of a subtropical climate for optimal growth of the coffee plant led to the conversion of primary rainforests in upland settings into shade-filled coffee groves.[40] Even when cultivators converted old growth forests into shade-filled coffee groves, the conversion led to degraded forests and a considerable loss of stored carbon. Because mature secondary forests in Guatemala contain more carbon than do coffee groves in the same region, the conversion of secondary forests into coffee agroforests most likely entails net losses of carbon in these landscapes.[41] A similar process of environmental degradation unfolded during the 1990s when agrarian reforms opened up the wooded central highlands of Vietnam for settlement and coffee cultivation.[42] Data on the emissions implications of the forest to coffee grove conversion do not exist, but it seems highly likely that most new coffee plantings, like cocoa plantings, degrade forests. Given these patterns, the long-term historical expansion in the extent of coffee cultivation implied continuing environmental degradation.

Amid the growing pressures on tropical forests, the increasing consumer demand for many of these tree products, including coffee, precipitated a wave of efforts after 1990 to make agroforestry more environmentally benign through ecocertification programs that linked agroforesters and environmentally concerned consumers. Cultivators would agree to practices that bolstered environmental services and to terms of trade that helped poor growers. The certifying organization would then send representatives to the farm to verify that the landowner had complied with the specified practices. The required practices almost always stipulated shade-grown rather than sun-filled fields, so certified producers typically sequestered more carbon than did uncertified producers with sun-filled fields.

Agricultural intensification frequently does not generate a corresponding decline in the prices of agroforestry crops such as cocoa, coffee, and rubber, because consumers react to the increased supply of the fruit by increasing their demand for it. For this reason, agricultural intensification in agroforestry does not generate land sparing. Given the elastic consumer demand, intensified production triggers a rebound effect in which farmers extend the cultivated area in response to the increased demand.[43] Under these circumstances, the optimal path to enhanced environmental services in agroforestry will most likely occur through ecocertification. As the history of the Roundtable on Sustainable Oil Palm (RSOP), recounted in chapter 5, illustrates, ecocertification in its initial form fell short of expectations. Still, growers, regulators, and consumers found much to like about the idea in its original conception, so it spread to the producers of other tropical crops.

Table 7.3 testifies to the attractiveness of the idea. It charts the growth across tropical commodities in the percentage of cultivators who have participated in certification programs. Waves of certification swept across some regions but not others after 2008, so certifiers have had uneven regional impacts. Three-quarters of the world's certified bananas grow in Ecuador, Colombia, and the Central American countries. While the Americas produced 52 percent of the world's coffee, they grew 77 percent of the world's sustainably produced coffee.[44] The uneven regional clustering of ecocertified producers suggests that momentum characterizes the adoption of certified agroforestry practices. Smallholders see their neighbors adopting sustainable practices and do likewise. Merchants in one place, but not another, join networks that market sustainable products. The same regionally uneven levels of engagement emerge among the certifying organizations.

Careful comparisons of certified and noncertified coffee cultivators in the province of Santander in Colombia make it possible to identify the effects of certification on coffee cultivation. Ninety-seven percent of the coffee farmers in this region work farms of less than five hectares. The farms typically contain a mosaic of irregularly shaped coffee groves of different ages. Certified growers are more likely to be full-time farmers than noncertified growers. Full time growers, with a stronger commitment to the agricultural enterprise, may have a longer time horizon and, for that reason, a greater interest in sustainable agriculture. More certified growers identified coffee as their chief source of income, and they devoted a larger

TABLE 7.3 Waves of Certification (Percent of Participating Cultivators: Across Crops and Over Time)

	2008	2012
Coffee	15%	40%
Cocoa	3%	22%
Palm oil	2%	15%
Tea	6%	12%
Bananas	2%	3%

Sources: Jason Potts et al., "The State of Sustainability Initiatives Review 2014: Standards and the Green Economy." International Institute for Sustainable Development (IISD) and the International Institute for Environment and Development (IIED), 2014, https://www.iisd.org /publications/state-sustainability-initiatives-review-2014-standards -and-green-economy; Ruth DeFries et al., "Is Voluntary Certification of Tropical Agricultural Commodities Achieving Sustainability Goals for Small-Scale Producers? A Review of the Evidence," *Environmental Research Letters* 12 (2017): 033001.

proportion of their land to coffee cultivation. Certified growers were more likely to plant trees on their farms outside of their coffee groves, and they planted a wider diversity of trees on their farms than did the noncertified growers. The price premium obtained for the certified coffee attracted uncertified growers initially. Once they joined the certification program, growers came to appreciate it because the program made them aware of new techniques for coffee cultivation.

The price differential between uncertified and certified Colombian coffee declined to 2 percent without the certification program losing significant numbers of members. Farmers did appreciate the less volatile price movements of coffee in the certified markets.[45] The persistence of the certified cultivators in the Colombian program, despite the decline in the price differential, has positive implications for the ability of certification programs to enlist and retain smallholders in their programs. The linking of payments for environmental services to certification programs would increase the appeal of the program to coffee cultivators. The shade trees in the canopy would earn the cultivator a stream of payments that would augment her or his income gains from the sale of high-quality coffee.

A meta-analysis of the effects of certification on the cultivation of tropical agricultural commodities (bananas, cocoa, coffee, oil palm, tea) indicates that certification has had positive environmental effects on the cultivation of these crops.[46] While each of these crops grows in the tropics, they get sold primarily in overseas markets. The market for certified coffee dwarfs the markets for all other certified commodities. Coffee certifiers represented 83 percent of the certification schemes in the analysis. The certified growers did engage in significantly more habitat conservation than did the noncertified growers. These findings indicate that certification programs, as least those focused on coffee, have produced real changes in growers' practices.[47]

The eventual environmental impact of certification programs on agroforests and plantations will depend on three variables: changes in the amount of unexploited forest in a region, changes in the substance of the certificates, and changes in the popularity of certified consumer goods. First, when primary rainforests disappear in a region, livelihoods based on accumulated soil fertility in forests, the forest rent, become less possible. It becomes less possible to earn a livelihood from serial relocations onto soils with accumulated fertility. Second, inclusion in certificates of absolute prohibitions on the conversion of primary forests into certified agroforests would discourage the most environmentally damaging of the conversions of forests into agroforests. Third, large numbers of consumers in lower middle-income countries such as China do not feel sufficiently concerned about the enhanced sustainability of certified coffee to pay the small increment in price associated with the certified product. As a result, smallholders currently supply the international market with more standard compliant certified product than merchants can sell at an appreciated price. For example, while 40 percent of all coffee comes to market, having been cultivated in compliance with certification standards, only 12 percent of the coffee gets sold in the more expensive certified markets.[48] Only growth in the popularity of certified products will produce the increases in prices necessary to make these environmentally friendly regimens sustainable.

A growing scarcity of unoccupied land, stricter qualifications for certification, and growing demand for certified products could together make certification an effective tool for restoring and expanding agroforests. Eco-certification schemes promise, like other corporatist arrangements, to build links between disparate sectors of society: producers open to conservation and distant consumers who want to purchase sustainably produced goods. REDD+ subsidies would make these links more robust by boosting the returns to producers whose agroforests sequester substantial amounts of carbon. Successful certification schemes become part of societal corporatist structures that organize disconnected producers, consumers, agribusinesses, and NGOs into a sector whose politics pivot around a common concern with natural climate solutions and sustainability.

Domestic Forests and Tree-Planting Campaigns

Domestic forests grow up in close proximity to human households, often just steps away from the back doors of houses. The location of the trees suggests the close connection that develops between the fluctuating needs of smallholder households and the plantings that over time shape domestic forests.[49] To one observer, "domestic forests" are "home gardens." To another observer, they are "kitchen forests." The interchangeable terminology signals the tight coupling between households, agriculture, and forests in these domestic settings. Crops and tree crops grow in close proximity to one another. Household members plant both of them. Under these circumstances, changes in the composition of domestic forests often reflect changes in the fortunes of the households who manage the forests.

The landholdings with domestic forests are usually small, often only several hectares in extent. In these settings, labor is cheap, and land is expensive.[50] Smallholders both practice agriculture and manage groves of trees on their landholdings, so changes in forest extent often link tightly to shifts in agricultural practices that in turn may respond to changes in household circumstances. The bonds between forests and households make it appropriate to call them domestic forests. Labor is relatively plentiful and poorly remunerated in places with domestic forests and other types of agroforests. These forests come in various forms. They range from "kitchen" forests of fruit trees outside the back doors of houses to small plantations of fast-growing exotics used for fuelwood or house construction.

Who does the planting varies according to the land tenure of a parcel. Where tenure is clouded or insecure, the occupants of a parcel often do not plant. Renters, sharecroppers, and tenant farmers do not plant trees. Wealthier households, with access to more land, plant more trees. Planted trees, in particular fast-growing exotics such as eucalyptus, represent an investment. Densely populated communities tend to foster more tree planting, in part because the large number of households provide ready markets for wood products such as the poles used in housing construction. Arid climates also spur tree planting because natural regeneration occurs sporadically, if at all, in dry climates. In arid settings, only domestic forests, where the families of smallholders periodically water seedlings, have trees with high survival rates.[51]

Planting varies with the gender of a smallholder. Women plant fewer trees when they do not have secure claims to land, as occurs quite frequently in patrilineal living arrangements where they move into their spouse's family compound. Female-headed households plant more trees, presumably because they have more secure claims to land than they do when they are not the head of a household. Men and woman tend to plant different sorts of trees. Women tend to plant fruit trees, perhaps because they contribute food to the household.[52] With the primary responsibility for collecting firewood each day, women appreciate the practical utility of planting trees near the house.[53] Nearby woodlots reduce the amount of time that women must spend collecting firewood each day. The dynamic of planting and harvesting in domestic forests implies that they only emerge and grow when smallholder households make long-term commitments to homesteads.

States and NGOs have attempted to reinforce tree planting in domestic forests with campaigns of collective action. In the 1970s, the late Wangari Maathai, a political activist in Kenya and later a Nobel laureate, mobilized groups of women in East Africa into the Green Belt Movement (GBM).[54] The GBM organized a succession of tree planting campaigns, each one scaled up in terms of its goals from the previous campaign. Most recently, global environmental NGOs launched a worldwide, trillion tree planting campaign. States, in addition to NGOs, have launched ambitious tree-planting campaigns. After announcing a goal of four billion planted trees, Ethiopia carried out a massive tree-planting campaign in 2019–2020 in which citizen participants planted 350 million trees.[55]

Since the 1970s, NGOs and states have launched tree-planting campaigns that have fallen short of achieving their goals because relatively few of the planted seedlings have survived. Critics of tree-planting programs have emphasized that the campaigns should focus more on tree growing than tree planting. The campaigns and participants could reduce planted tree mortality significantly by paying more attention to the management of the newly planted trees. Effective management of the planted seedlings would become more likely if campaigns ceded control over the new plantings to local communities with long-term interests in the trees and forests.[56] Short-term interventions do not work. To bolster seedling survival rates, individuals and communities would have to commit themselves to five years of work to establish a forest. In effect, tree-planting campaigns, to succeed, require a commitment to the planted forests commensurate to the commitments made by smallholders to the domestic forests around their houses.

The historical narrative surrounding the emergence of domestic forests became much clearer during the early 1990s with the publication of a book-length case study, entitled *More People, Less Erosion*,[57] about the expansion of domestic forests in the Machakos district of Kenya between the 1930s and the 1980s. How extraordinary were the conditions that precipitated the expansion of the domestic forests in Machakos? To answer this question, researchers at the Overseas Development Institute carried out a comparative historical study of six similar cases elsewhere in sub-Saharan Africa during the late 1990s. They did not find a general trend towards environmental recovery. Three conditions did seem essential for expanding domestic forests. (1) Smallholders maintained a primary focus on earning their livelihoods from agriculture. (2) Productive land remained scarce, and (3) planting trees offered a clear way to increase revenues for smallholders.[58] These conditions presume secure land tenure. It is a necessary but not sufficient condition for the daily investments of labor by smallholders that make domestic forests grow.

Environmental Degradation, Societal Corporatism, and Restoring Agroforests

For centuries, Indigenous peoples cleared and planted new sites for crops and fruit trees every few years.[59] Since the 1970s, many shifting cultivators have abandoned this routine. They intensified agriculture in a single place.[60] With intensification, they became more sedentarized and produced cash crops from trees and shrubs such as oil palm, rubber, coffee, cocoa, and damar for decades on end. Despite the economic disadvantages faced by smallholder agroforesters, large landowners have not typically been able to dislodge smallholders from their lands because smallholders produce such large harvests per unit area.[61]

Small-scale agroforesters are not disappearing. They can absorb the extra costs of planting cash crops because members of the family usually do the extra work. These household members are typically more skilled and poorly paid than hired workers from outside the family.[62] Because agroforesters produce a diverse mix of crops in small volumes, they face diseconomies of scale when they sell their harvests. Again, family members provide the poorly compensated labor that get these products to market. Depending on the crop, smallholders may supply processors on large plantations with product for their mills for lengthy periods.[63] Smallholders have established ties to these mills or to neighboring large landowners to insure that they get seeds, technical assistance, and purchasers for their product when they harvest it.[64] These ties between large plantations with mills and surrounding smallholders repeat themselves across different agricultural sectors. They have strengthened smallholder participation in both the rubber and oil palm agricultural sectors in Indonesia. To increase their income still further, many smallholders have again intensified their agriculture, creating small plantations of either monoclonal rubber or oil palms on two or three hectares of land. These land uses no longer resemble agroforests. With frequent clearings and reclearings of the land, these plantations probably emit more carbon than they sequester.

In rubber tapping, coffee, and cocoa-producing districts, large numbers of smallholders have continued to practice long established shade-filled coffee, jungle rubber, and shade-filled cocoa regimens of cultivation. Smallholders have increased the profitability of these practices through the selective adoption of technological innovations such as, for cocoa growers, rubber tree canopies and new hybrid varieties of seeds. Because these agricultural regimens sequester more carbon than sun-filled plantation-like agriculture, these long-established agroforests exemplify the kind of landscapes that climate conscious societies would want to promote.

Observers find much to like about agroforestry as a land use because it promises both intensified agriculture and expanded forests in a world facing chronic food insecurity and destructive climate change. The recent history of agroforestry underscores the difficulties of upscaling agroforestry to meet the challenges of climate change and rural poverty. Jungle rubber, cocoa, and coffee all triggered extensive processes of deforestation and environmental deterioration when they first spread across forested regions of the tropics. Only when environmental deterioration closed off opportunities for further agricultural expansion did the possibilities for carbon sequestration through the rehabilitation of degraded jungle rubber forests or old cocoa groves begin to increase.[65] In each of the three cases reviewed here, rubber, cocoa, and coffee, the environmental degradation that accompanied the expansion of agricultural production has spurred efforts to revive more forest-friendly agroforestry practices.

Ecocertification programs offer a resurgent carbon-sequestering path to an expanded agroforestry sector. Since the 1990s, NGO activists have created ecocertification initiatives for every major tropical crop (see table 7.3). The environmental effectiveness of these NGO-led programs has varied, with coffee certification more effective, cocoa less so, and rubber least so. These certifying organizations provide focal points for additional efforts to strengthen the environmental services provided by coffee, cocoa, and rubber producers. Payments for environmental services could certainly become an important element in this mix of ecocertified initiatives. More concerted consumer buy-ins to certified products, already considerable in the case of coffee, would increase the contributions of ecocertification programs to natural climate solutions.

If agroforestry is to become a viable option for reducing hunger and limiting the impact of climate change, state services to rural people must improve. Successful ecocertification schemes and payments for environmental services programs, like the expansion of domestic forests, all require formal arrangements that recognize the ownership of small landholdings in rural and periurban places. Agricultural development agencies in Ghana, Côte d'Ivoire, and Indonesia promoted the planting of new hybrid variety cocoa seedlings. They exemplify the improvements in government support for agroforestry. In these instances, the potential of agroforestry to bolster environmental services such as carbon sequestration depends on the ability and willingness of state actors to build institutions that allow smallholders to reap predictable rewards from their labor on the land.

Table 7.4 sketches out the changing human ecology of the global agroforestry sector. Political pressures from smallholders, as well as NGO activists, can contribute to the necessary strengthening of state and third sector (NGO) institutions in agroforestry. Politically active smallholders, NGO certifiers, public officials, and affluent, environmentally concerned

TABLE 7.4 The Human Ecology of Agroforests, Post-2000

	Jungle rubber	Shade-filled cocoa	Shade-filled coffee	Domestic or kitchen forests
Biodiversity	High	High	High	High
Land tenure security	Low	Low	High	High
Size of landholdings (in hectares)	2–10	~3	2–10	1–2
Certifiers' institutional strength	Weak	Weak	Strong	Weak
State programs to strengthen agroforestry	No	Yes	Yes	Yes
Trends in carbon-sequestering land uses	Contracting	Unclear	Expanding	Expanding

consumers would together comprise the basic elements of a societal corporatist organization capable of creating and implementing a set of resurgent carbon-sequestering agroforestry programs. Ecocertification programs would become, in this setting, important building blocks in the construction of a societal corporatist political order that would implement natural climate solutions.

8

• • • •

Resurgent Forests

A QUALITATIVE COMPARATIVE ANALYSIS

Analyses from the collected case studies in the five previous chapters provide a way to assess the argument that corporatist political processes have facilitated the expansion of forests in disparate locales around the globe since the 1980s. The cases range across a wide variety of landscapes: intact forests, abandoned agricultural lands, recently created forest plantations, silvopastures, and agroforests. Table 8.1 lists when and where these episodes of forest conservation, restoration, and expansion occurred. Twelve of the nineteen case studies focus on recent changes in forest cover in the Global South. In this respect, the geographical distribution of the case studies reflects our ongoing preoccupation with the rapid deforestation occurring in the Global South and the recognition that natural climate solutions must entail an end to deforestation. Three of the landscape transformations described in the case studies occurred in the twentieth century. All of the others began in the twentieth century and extended into the twenty-first century. Forests increased in extent in twelve of the nineteen cases. The validity of the argument about the pivotal role of corporatist political processes in forest recoveries will be clearer if the analysis begins with a brief description of the broader historical context within which the recoveries occurred.

TABLE 8.1 Case Studies of Forest Gains and Losses

Types	Time period	Forest cover trends
Avoided Deforestation		
Ecuadorian Andes–Amazon	1975–2014	Continued old growth in parks ↑
Secondary Forest Expansion		
New England, United States	1840–1950	Upland reforestation ↑
NW Portugal	1964–1995	Dry land reforestation ↑
New Deal South, United States	1935–1975	Eroded land reforestation ↑
Northern Costa Rica	1985–2005	Upland pasture reforestation ↑
Forest Plantations		
Congo	2000–2020	Concessions: logging and regrowth
China	1991–2010	Smallholder plantation ↑
Vietnam	1992–2005	Smallholder plantation ↑
Laos	2000–2020	Concessions, little reforestation
Indonesia	1998–2012	Intensified agriculture, little reforestation
Southeastern United States	1980–2017	More forest plantations
Chile	1980–2011	More forest plantations ↑
Silvopastures		
Upper Amazon	1985–2015	Spontaneous silvopastures ↑
Lower Amazon	1980–2000	Class conflict, little reforestation
Niger, Sahel	1985–2010	Planted trees ↑
Agroforests		
Rubber	1990–2020	Intensified agriculture, little reforestation
Cacao	2010–2020	Intensified agriculture, little reforestation
Coffee	2005–2020	Ecocertified forest resurgence ↑
Domestic forests	1990–2020	Tree-planting campaigns ↑

Note: The ↑ symbol indicates that significant increments in carbon sequestration and other environmental services are occurring in these landscapes.

Sources: Case studies in chapter; see notes.

An Explanation: The Great Acceleration, Corporatism, and Forest Resurgence

Forest Resurgence in the Aftermath of the Great Acceleration

The years since the end of World War II have witnessed a great acceleration in the scale of the human enterprise.[1] Its impact on forest cover has been unmistakable. To feed and shelter expanding human communities, coalitions of farmers, government officials, and agribusinesses have converted millions of hectares of forest into croplands. These business-as-usual growth coalitions have transformed the global landscape. Burgeoning demand from growing numbers of affluent consumers have driven up the prices for wood and agricultural commodities. The high prices have in turn incentivized land clearing by growth coalitions in the Americas, Asia, and Africa.

The participants in these business-as-usual coalitions have varied across the regions. In West Africa, smallholders did almost all of the land clearing. Distinctions got drawn between different groups of land clearing smallholders. Some were migrants from the Sahel. Others were long-term residents of the cocoa belt. In Laos, Chinese and Vietnamese companies sought to purchase or lease degraded lowland forests. They planned to convert these lands into highly productive rubber plantations. On the Indonesian island of Kalimantan, oil palm companies claimed and cleared large tracts of old growth forests, planted oil palms on the cleared land, and recruited smallholders to plant more oil palms on adjacent lands, with contractual understandings that the smallholders would process their fruit at the companies' mills. Politicians from local, state, and national governments played important roles in the legitimization of the land grabbing that occurred simultaneously with the destruction of the forests.

The salience of these business-as-usual coalitions in forest destruction explains, obliquely, why so much reforestation has occurred on degraded lands since the beginning of the nineteenth century. Corporate representatives, ambitious smallholders, and enterprising politicians, seeing little of value in these degraded, cutover lands, have left the old settlement zones. With stocks of above and below ground carbon depleted in these places by earlier rounds of exploitation, new land uses such as forest plantations, while not generating unusually large stores of carbon, still on net, increased the stores of carbon because the previous, degraded land use, such as an old cattle pasture or an invasive infested old field, stored so little carbon. Proposals for forest restoration on these sites do not face competing proposals for land use because the degraded sites offered few other economic opportunities.

The spread of silvopastures in old settlement zones of the Ecuadorian Amazon, described in detail in chapter 6, approximated this restoration dynamic. Globalization, abetted by road construction, spurred extensive destruction of forests. Cattle ranchers cleared away the forests and planted pastures. Three decades later, the land clearing had ended. The pastures had grown old, with reduced rates of regrowth, so they supported smaller numbers of cattle. In this context, small-scale cattle ranchers sought to augment their incomes by allowing commercially valuable trees to regenerate in their pastures. Modest increases in sequestered carbon occurred as many pastures became silvopastures. In another variant of the same dynamic, loggers in riverine forests in the lower Amazon cut the large old growth trees and then moved on to cut down other old growth forests in nearby regions. *Caboclos*, long resident along the Amazon, stayed in the logged over region, harvested a wider range of trees, and promoted the growth of acai agroforests whose fruits they sold in the urban markets of Belem. The growth of these secondary forests increased the carbon stored in the logged over lands.[2]

In other instances, labor market dynamics in rapidly growing economies precipitated land abandonment and reforestation. In New England during the closing decades of the nineteenth century, young women left exhausted lands on upland farms to work in recently established textile mills in nearby towns. Their departures, coupled with the declining fertility of the old fields, precipitated land abandonment and forest regeneration. In a similar instance in northwestern Portugal, the easing of restrictions on foreign workers in wealthy European Union polities such as France precipitated a wave of departures from farms in Portugal as workers found that they could make much more in wages in French cities than they could cultivating small, water-deficient fields on Portuguese farms.[3] In both cases, the declines in the availability of labor, coupled with declining yields from old fields, caused land abandonment and the reemergence of forests on the abandoned land. Reforestation occurred after people had destroyed old growth forests and depleted the soils of the deforested lands.

The Countermovement, Corporatism, and Resurgent Forests

The loss of so many forests and the onset of global warming in the late twentieth century alarmed observers, and they began to agitate for changes to counter the destruction. Beginning in the 1980s, ecologists, Indigenous

peoples, and activists from environmental NGOs mobilized a countermovement dedicated to the preservation, restoration, and expansion of forest
cover. In effect, the forest sector of the expanding global economy experienced what Karl Polanyi called a "double movement."[4] The forest-destroying
Great Acceleration constituted the first movement. It provoked a second,
countermovement dedicated to saving and restoring the forests. The ambitions of the activists in the countermovement were global in scale, but their
ability to effect change remained open to question. This book addresses this
question through a series of case studies that together suggest an explanation for the variable outcomes of recent efforts by activists at forest conservation, restoration, and expansion.

In a line of reasoning presented in chapter 2, the countermovement has
enabled a resurgence of forests in places where activists have constructed
corporatist political processes. Corporatism (1) organizes political processes
by sectors of the economy, so in this instance a network of parties interested in agriculture and forests would come together to discuss, enact, and
implement policies pertaining to landscapes. The network includes (2) a
diverse collection of interested parties, ranging from wealthy global elites
to impoverished Indigenous peoples. These parties have ties to both global
and local interests. These varied interests (3) negotiate with one another
and arrive at compacts that bind the signatories to courses of action in the
sector over delimited periods of time.[5] The compacts (4) articulate new,
forest-enhancing norms about the use of land. They also commit the parties (5) to particular courses of action that implement the new, forest-
promoting norms.

The tasks of the different parties to the compacts reflect their social
structural positions. The scientists identify how to restore forests. The global
interests, aided and abetted by nongovernmental organization (NGO) activists, provide financial support for the acquisition of forests or for carbon
sequestration payments. Local landowners promote tree growth on their
land. Government agents and NGO personnel provide the institutional
supports, such as secure land tenure and government subsidies, that facilitate forest growth and regrowth. The transfers of funds for forest restoration from the center to the periphery of the world economy have an overall
redistributive effect. The substance of the sector's commitments varies

across different types of landscapes. For example, the expansion of a protected area of intact forest may entail payments by donors to purchase tracts
of land. In other instances, wealthy overseas consumers pay extra for an
ecocertified commodity such as organic coffee.

The ties between these disparate parties bind because the countermovement strengthens the resolve of all the parties to the compact. It becomes
the vehicle through which the different parties get something done in their
fight to restore or enhance forests. Activities that erode the solidarity of the
movement, such as corruption or a failure to deliver on promises, lessen
the political force of the corporatist process. Only where corporatist processes provide a kind of political glue will climate stabilization coalitions
accomplish a collective end such as forest restoration. Even then, the ability of corporatist political processes to foster forest recovery depends, in
part, on the capacity of a corporate grouping to forestall further forest
destroying activities by business-as-usual growth coalitions.

Corporatism and Reforestation: A Qualitative
Comparative Analysis

Does this account of forest resurgence have an empirical basis? The case
studies answer this empirical question through detailed narratives about
the ways in which individuals, communities, and companies came to add,
preserve, or restore specific forests. A qualitative comparative analysis across
the nineteen case studies assembled here provides a broader scale and more
definitive answer to this question. What kinds of events or conditions associate across the cases with forest expansion, conservation, or restoration?

Qualitative comparative analyses (QCA) use set theory to identify particular conjunctures of causes that produce an outcome of interest.[6] The
frequency with which a particular conjuncture or set of circumstances associates with an outcome of interest suggests but does not prove a causal
connection between a conjuncture of conditions and an outcome of interest. In our analysis, the outcome of interest is carbon sequestering reforestation. The argument outlined in chapter two identifies corporatist political

processes, with its participants, as a crucial facilitating factor in contemporary efforts to foster increases in forest cover.

The histories of individual efforts to reforest land suggest the following influential factors in the shorthand used in the equations in box 8.1: *time*—when the reforestation took place; *mig*—rural to urban migration from a reforesting community; *inter*—involvement of international institutions in efforts to foster forest resurgence; *ngo*—the involvement of environmental or Indigenous advocacy groups in a forestation effort; *gov*—active participation of government officials in forestation efforts; *landholders*—the large or small landowners whose tracts of land undergo forestation; *corpor*—a political process that brings together the interested parties to negotiate compacts that guide the forestation efforts. Uppercase letters indicate the presence of a factor in a particular case. Lowercase letters indicate the absence of a factor in a case.

Box 8.1 presents the results from a qualitative comparative analysis of causal factors in the nineteen cases described in chapters three through seven and listed in box 8.1. The association of corporatist arrangements with forest resurgence has occurred across a wide range of landscapes. The first term in box 8.1, "INTER NGO GOV LANDHOLDERS CORPOR," associates with forest resurgence across five different cases. Each case involved a diverse mix of landowners (LANDOWNERS), activists (NGO), government officials (GOV), and representatives from international organizations (INTER), all of whom came together in corporatist, forest-saving coalitions.

The wide range of social and ecological circumstances in which corporatist processes fostered forest regrowth is striking. Cattle ranchers in Costa Rica, Indigenous peoples in the Ecuadorian Amazon, peasants in arid zones of the Sahel, coffee cultivators in Colombia, and activists in tree planting campaigns all joined corporatist coalitions whose members made different kinds of contributions to restoration or conservation efforts. The types of corporatist-influenced forest resurgence varied as well. Corporatist decision making has fostered forest preservation (Ecuador), secondary forest expansion (Costa Rica), the emergence of silvopastures in arid landscapes (Niger), and the reinvigoration of coffee agroforests in humid forests (Colombia).

BOX 8.1

A Qualitative Comparative Analysis of Forest Gains

INTER NGO GOV LANDHOLDERS CORPOR (5) +

inter ngo GOV LANDHOLDERS CORPOR (2) +

time ngo gov MIG LANDHOLDERS corpor (2) +

time NGO GOV MIG LANDHOLDERS CORPOR (1) +

NGO GOV LANDHOLDERS CORPOR (1) +

ngo gov LANDHOLDERS corpor (1)

Interpretation: Lowercase expressions mean that that the group did not participate in the process that spurred forest expansion. Uppercase expressions mean that the group or aggregate of people did participate actively in the reforestation process. The numbers in parentheses indicate the number of cases that fit that particular configuration of causes.

Factors: INTER = United Nations, World Bank, affiliated global institutions; NGO = nongovernmental organizations, local or national (usually environmental or Indigenous); GOV = government officials whose work pertains to landscapes, agriculture, or forests; MIG = rural–urban migrants who left the region; LANDHOLDERS = the farmers or other individuals who work the land; TIME = historical period; CORPOR = presence of a political process that brings together representatives from centers of power and local landowners to negotiate compacts that preserve or restore forests.

Proponents of REDD+ schemes have also set up corporatist processes and used them to preserve old growth forests.[7] Corporatist arrangements have been used just as often to facilitate tree planting.[8]

All of the cases in the first causal configuration in the QCA are instances of societal corporatism. For example, in Costa Rica, a parliamentary democracy, trends in international political economy, tourism receipts, and beef exports persuaded legislators to reduce economic protections for cattle ranchers during the 1980s and 1990s. This shift in agricultural policy promoted the expansion of secondary forests into upland areas of northern Costa Rica.[9] The wide range of these situations in which societal corporatist processes have contributed to forest gains suggests their overall importance in the resurgence of forests.

The second term, "inter ngo GOV LANDHOLDERS CORPOR," captures the experience of China and Vietnam with reforestation since 1990. In China, crisis narratives spread among party-dominated cadres of leaders in the aftermath of natural disasters. Compelling accounts about the ways in which denuded, upland landscapes contributed to deadly downstream flooding strengthened the commitment to the reforestation program of high officials, local officials, and smallholders. The degraded state of Vietnam's forests, coupled with the disastrous events next door in China, convinced Vietnamese officials to implement a reforestation program. Difficulties in implementation tested the strength of these commitments in China. When the initial plans proved inadequate, the central government as well as local authorities persisted with revised plans. Smallholders accepted the government-provided seedlings, but when these plantings failed, household heads and local authorities searched for substitute plantings. Public officials continued to promote the program amid these difficulties. In China, the revised plans, new initiatives, and sources of additional support came from local officials as well as from the rulers of the central state.

The third term in the equation, "time ngo gov MIG LANDHOLDERS corpor" captures the experience of two earlier episodes of reforestation, in the northern Appalachian Mountains of the United States during the first half of twentieth century and in northern Portugal during the second half of the twentieth century. In both instances, out-migration of small farmers led to the abandonment of degraded, difficult-to-farm lands. Naturally regenerated secondary forests soon occupied the abandoned land. Corporatist processes did not play a role in these processes of reforestation. This circumstance suggests that the link between corporatist organization and forestation has historical limits. It has become more common with the growth of state and NGO power since the 1970s.

The fourth term in the equation, 'time NGO GOV MIG LANDHOLDERS CORPOR' captures the New Deal experience in the southern cotton belt of the United States. While this region saw the same pattern of rural to urban outmigration that occurred earlier in the northern Appalachians and later in northern Portugal, public officials, as well as spun-off, farmer-led, soil conservation service organizations, induced farmers to cut back on the acreage that they farmed. These lands went back into naturally regenerated secondary forests, beginning during the 1930s.

Roosevelt's "brain trust," soil scientists, and farmer-led conservation groups created a corporatist process for converting farmlands back into forests. In its organizational structure, the New Deal prefigures contemporary efforts to restore forest lands. These historical similarities suggest that a widespread, contemporary resort to natural climate solutions could be fairly construed as part of a Green New Deal.

The next term in the equation "NGO GOV LANDHOLDERS CORPOR" captures Chile's recent experience with reforestation. The most salient aspect of the Chilean experience concerns the prominence of private timber companies in the state's plan to reforest a significant proportion of the Chilean landscape. Large timber companies in other places such as the United States have not taken such a prominent role in reforestation efforts. NGOs have also played a prominent role in Chilean planning for the forest sector. As typically occurs in corporatist processes, the NGOs have negotiated directly with the timber companies. The Chilean experience underlines a potential role for large companies in future efforts to restore or expand forests.

The final causal configuration, "ngo gov LANDHOLDERS corpor," explains the spontaneous emergence of a silvopastoral landscape in a cattle ranching region of the Ecuadorian Amazon. In a context of declining soil fertility in forty-year old pastures, small-scale cattle ranchers allowed sprouting, commercially valuable tree seedlings in pastures to grow to full size. After fifteen years of growth, the lumber from these trees could be sold to sawmills, thereby creating secondary streams of income for the ranchers. The shade-filled pastures continued to provide forage for cattle at roughly the same rates as sun-filled pastures. The spread of seedlings from pasture to pasture occurred spontaneously, without promotion by government or NGO personnel. The absence of NGO or state assistance, despite the incipient provision of environmental services, recalls the earlier twentieth-century experience with spontaneously regenerated secondary forests in the northern Appalachians and Portugal. These three cases underscore the contingent nature of government engagement with reforestation projects. Quite frequently, governments do not provide meaningful support for reforestation.

If forest gains do somehow occur in this context, they may well have been smaller in scale than the gains that would have occurred with government support. If this conjecture is correct, it leads to the conclusion that the corporatist processes that advance reforestation efforts do not just happen. In most cases, they must be promoted and constructed by committed political leaders with ties to a social movement.

Looking across all of the configurations of causal conditions in the QCA of natural climate solutions, the association of corporatism (CORPOR) with successful instances of forestation is strong. Forest cover has increased in places with societal and state corporatism. It has not increased in places where states and governance have been weak or where crucial actors such as state officials or local landholders have not become participants in sectoral agreements to preserve or restore forests. The case studies provide many examples of these forest-destroying patterns. The business-as-usual coalitions of corporate planters, hacienda owners, small landholders, and bankers have worked relentlessly to convert forests, either directly or indirectly, into agricultural commodities with help from politicians with prior commitments to forest protection.[10] When and where opposition to the land clearing has surfaced, pro-growth politicians with ties to business-as-usual interests have used their influence to prevent the protection of forests. As a result, efforts to prevent the destruction of forests through the creation of preserves, like those in old growth forests of lowland Kalimantan, have sometimes proved ineffectual.[11] Undoubtedly, a high incidence of corruption in the public sector has made it virtually impossible to negotiate the binding, corporatist compacts that avoid deforestation and restore degraded forests.

Class conflict also crippled corporatist efforts at conservation. The case study of babassu silvopastoral landscapes in eastern Amazonia exemplifies this dynamic. The struggle between large-scale ranchers and landless peasants who wanted to collect the fruits from babassu palms on the ranchers' pastures effectively prevented the expansion of the babassu silvopastures. The ranchers chopped down the babassu trees and planted pastures that discouraged the spread of babassu, all in an attempt to prevent local inhabitants from entering the ranchers' lands to harvest the fruits. More generally, corporatist processes added political value when they brought together diverse parties to negotiate compacts about forests. This organizational advantage disappears in an atmosphere riven by class conflict where the parties do not trust one another. In this kind of atmosphere it becomes impossible to negotiate corporatist compacts between large and small landholders.

The other, related pattern running through the case studies entails an association between good governance and forest cover gains. Corporatist political processes have historically taken root in polities with strong states, in other words, in states where rulers enjoyed legitimacy, citizens participated in the polity in large numbers, and states could mobilize people to pay taxes and fight wars.[12] Countermovements such as the forest-enhancing conservation movement have probably strengthened states by increasing citizen participation in public affairs. Movement activists could through surveillance deter cheating by signatories to the compacts. In this manner, growth in the climate stabilization movement should make for more effective compacts and increase the likelihood of workable corporatist arrangements, even in places long characterized by weak governance.

Regional Patterns of Reforestation

Constructing corporatist political processes to enhance forests would almost inevitably begin in diverse ways given the range of different settings in which forest restoration and expansion has occurred. These efforts might take, as a point of departure, recent changes in landscapes and try to build on these changes. David Kaimowitz, then the director of the Center for International Forestry Research (CIFOR), made this point succinctly. "To succeed, strategies should capitalize on the trends that are already driving the economy and people's decisions and nudge them in the right direction."[13] This operating principle of sustainable development, applied to efforts to expand forest cover, acknowledges the different historical circumstances in which forest cover has expanded during the twentieth and twenty-first centuries. These historical geographies had regional dimensions that followed the contours of terrain, climate, soils, and society.

These episodes of forest restoration and expansion, following Kaimowitz's suggestion, might provide templates for later forest gains that states, NGOs, and communities might pursue as part of a larger climate policy. In this respect, efforts at forest cover increases would take a "polycentric" form that is different from region to region and at the same time instructive about further forest-enhancing policies.

The leaders of the IPCC effort to minimize climate change adopted this polycentric approach to climate governance in 2015 when they called for each country to submit a plan for a Nationally Determined Contribution (NDCs) to the global mitigation effort. The planners in each country assessed how, given their human and natural resource endowments, they could cut their greenhouse gas emissions and/or sequester sufficient amounts of carbon to achieve carbon neutrality by a specified date (usually 2050 or 2060) in the future. Because resource endowments such as forests, winds, and flowing water vary regionally, with important implications for generating energy and sequestering carbon, plans for reducing emissions, outlined in the NDCs, also varied regionally. Accordingly, efforts to sequester more carbon through escalated efforts at forest restoration or expansion would vary with the type and extent of forest cover change across regions. Kaimowitz's point about the gains to be had in the efficacy of political pressure, when government and NGO initiatives build on local trends, becomes pertinent here. To maximize the impacts of overseas assistance, donor groups and representatives would engage with political leaders and local landholders in projects that reinforce initiatives that farmers have already begun. In other words, external agents would try to meet people where they live.[14]

To this end, corporatist political processes would vary across landscapes, acknowledging local features and trends in the coupled natural and human systems of each region. Mountainous terrain would matter, but so would the size of landholdings and trends in the extent of forest cover. Together these attributes create regional landscapes that in turn shape feasible strategies for promoting forest conservation and resurgence in each region. The following paragraphs outline eight regional landscapes and the corresponding political strategies that would promote forest conservation and resurgence in each of these regions. Figure 8.1 maps these regional patterns.

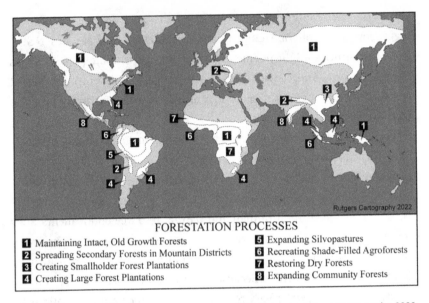

FIGURE 8.1 Map of regional forestation processes, 2020. Drawn by Rutgers Cartography, 2022.

Intact Subarctic and Tropical Forests
in Sparsely Populated Regions

Five nations contain large, intact forests. Three of these nations, Canada, Russia, and the United States, contain large, forested expanses of land in the subarctic zone. Two nations, Brazil and the Democratic Republic of the Congo, contain large, intact forests in the tropics.[15] The upper reaches of the Amazon basin in Bolivia, Peru, Ecuador, and Colombia contain intact extensions of the Brazilian Amazon forest. Melanesia, in particular the island of New Guinea, also has large blocks of intact forests that stretch down to the sea. Maintaining the existing forests, path #1 in chapter 2, seems the highest priority in these places. Two strategies, protected areas and Indigenous preserves, would appear optimal in these settings. The intact condition of the forests in these regions suggest that alternative economic opportunities in agriculture or mining have not been present in these regions, so the purchase of blocks of these forested lands for conservation might be possible. The Indigenous populations in these places have forest-based livelihoods, so they would be inclined to defend the forests. Historically, Indigenous peoples have had much more success acquiring control over forested lands in Asia and Latin America than they have had in sub-Saharan Africa, so the success of this strategy hinges on variations in local balances of power between commercial elites and Indigenous peoples.[16] Both the creation of nature reserves and Indigenous preserves would become more effective if REDD+ programs could provide payments for the varied environmental services provided by the old growth forests in these regions. Prototypes for these conservation-oriented REDD+ programs already exist, particularly in Brazil.[17]

Expanding Secondary Forests in Montane Landscapes with Declining Populations

Mountain ranges in diverse locales, the Himalaya in Asia, the mountains of Mesoamerica, the Andes in South America, and the Carpathians in Eastern Europe contain areas in which continued out-migration by smallholders has recently led to the abandonment and reforestation of agricultural lands. Declining yields from marginal agricultural lands and the prospect of better paying jobs in cities, oftentimes in other countries, has encouraged smallholders to leave these lands. Over time, these "agricultural adjustments" have redistributed forests toward the uplands and agriculture toward the lowlands.[18] This land abandonment dynamic characterized upland portions of Western Europe and the northeastern United States during the nineteenth and twentieth centuries. With few farmers left in these zones, it no longer does so.

Agricultural land abandonment has, however, become more prevalent in upland places in the Global South with persistently high population densities. In Nepal, for example, land abandonment has concentrated in the mountainous districts of the Himalaya and not in the Gangetic plains of the Terai district.[19] In many of these districts, the owners did not abandon the land outright. Instead, they planted trees. The resulting smallholder forest plantations required little labor from the owners, so labor

migrants could leave their villages of origin to work elsewhere without sac-
rificing the income earning potential of their newly planted woodlots.[20] In
their absence, the owners' investments in trees would grow until they
yielded a return at harvest time. Subsidies from governments for allowing
agricultural lands to lie fallow, similar in their operation to conservation-
oriented REDD+ programs, have facilitated the regeneration of secondary
forests in the United States and the European Community. Because sloped
land typically contains more ecological niches than flat land, the reversion
of sloped lands to forests has probably benefited biodiversity in these
locales.[21]

Smallholder Forest Plantations in Densely Populated Rural Districts in Asia

South, East, and Southeast Asia have witnessed substantial increases in the
numbers and extent of small, one- to five-hectare forest plantations since
1990. In mainland Southeast Asia, small, intensively cultivated plantations
have spread into adjacent upland areas. State officials played a central role
in the creation of these plantations. Transmigrant program officials in Indonesia promoted ties between oil palm plantation companies and transmigrant colonists that facilitated the spread of small plantations among the
colonists.[22] Chinese officials overcame high seedling mortality in implementing the "grain for green" program.[23] In both China and India, substantial numbers of afforestation and reforestation projects with exotic species
received REDD+ support.[24] Road building in both Vietnam and India facilitated the movement of wood products to urban markets. In so doing, the
road building has encouraged the creation of forest plantations in corridors along the new roads.[25]

The frequency of harvests from smallholder plantations, under current
practices, creates a substantial problem regarding payments for carbon
sequestration. The harvests of trees after only eight to fifteen years makes
it virtually impossible for these plantations to sequester enough carbon to
warrant carbon offset payments. Similarly, the rapid spread of rubber

monocultures onto terraced fields in adjacent upland areas of mainland Southeast Asia represents an extension of the intensive plantation monocultures typical of this region. Although rubber monocultures may be considered forests, their spread onto terraced fields brings negative environmental repercussions such as increased greenhouse gas emissions and biodiversity losses to the region.[26] Longer land use rotations that postpone harvests until forty years after planting might allow enough carbon sequestration in these "forests" to qualify them for payments for environmental services through REDD+ programs.

Large Corporate Forest Plantations in Humid Agricultural Zones

Timber companies harvested large volumes of trees from the humid forests of the Pacific Northwest in North America and Scandinavia in Europe throughout the twentieth century, and they sought still larger harvests of wood by planting monocultures of fast-growing, often exotic species of trees in humid regions during the late twentieth and twenty-first centuries. The owners of these new plantations pursued, above all, high profits. The locations of the plantations convey this emphasis on maximizing profits. In the southeastern United States, the companies established new pine plantations in close proximity to the cities of Atlanta and Charlotte that provided markets for the wood products of the plantations.[27] The planted forests in Chile and southern Brazil also grew in close proximity to major urban centers in both countries. Similarly, investors in the new, large-scale timber and oil palm plantations in insular Southeast Asia have located all of them in the coastal zones of Borneo and Sumatra.[28] Proximity to deepwater ports provides plantation owners with vital savings in transportation costs, given that they ship harvested wood overseas.

Despite the relatively rapid uptake of carbon by the fast-growing trees in these plantations, their utility as potential carbon sinks seems low because the harvests take place so frequently. The harvests themselves induce emissions, and the wood products become a continuing source of emissions after the harvests, depending on their postharvest use. Given the initial rapid uptake of carbon in these trees, the basis for a REDD+ protocol exists in these plantation belts, but only if the timber companies agree to a change in norms in which longer intervals of time between harvests become the standard practice.

Silvopastures Along the Edges of the Amazon

Through seemingly relentless processes of tropical deforestation since the 1970s, farmers, colonists, and investors have cleared almost half of the tropical biome of trees.[29] Pastures now predominate in many of these deforested places. The pastures stretch around the edges of the Amazon, eastward from the Andean mountains and northwestward from urban centers in Brazil. Pastures also predominate in Central America and southern Mexico. Seed rain from remnant forests in the humid portions of these pastoral landscapes has spurred the gradual repopulation of these pastures by trees, especially where shade-tolerant pasture grasses predominate. Farmers in these places do not uproot the seedlings of commercially valuable tree species when they first sprout in pastures. The seedlings promise to generate a stream of income ten to fifteen years later when, as young trees, they become large enough to sell to local sawmills.[30]

Tree densities in many of these pastures have increased to the point where they sequester significant amounts of carbon each year. In the Ecuadorian Amazon the emerging stands of trees in pastures sequester, on average, about one ton of carbon per hectare of pasture each year. These increments in sequestered carbon, coupled with the well-documented increases in the extent of silvopastures, make a case for REDD+ projects that focus on using payments for carbon sequestration to accelerate the conversion of pastures into silvopastures. If activists are serious about exploring every possibly fruitful avenue in a time of crisis, then calculations of small gains in sequestered carbon in silvopastures and payments for the sequestered carbon become justifiable.

Dense Agroforests in the Humid Tropics

The expansion of coffee and cocoa cultivation into belts of old growth forest in Southeast Asia, West Africa, and Latin America since the 1970s has degraded these forests. The forest losses have been especially large where norms about the desirability of sun-grown cultivation have spread among smallholders. Sun-grown coffee and cocoa both promise faster maturation, so growers do not have to wait for four years to harvest their first crop from sun-filled fields. The yields from sun-grown coffee and cocoa are also superior to shade-grown coffee and cocoa, but the length of time in which the sun exposed plants typically yield a harvest is much shorter (~15 years) than is the case with shade-grown plants (~30 years). Successive replantings of coffee or cocoa plants in sun-grown conditions further degrades the soils and promotes, at least in Indonesia, the spread of invasive grasses such as *Imperata cylindrica.*

Under these dire circumstances, some environmentalists have come to support the reestablishment of shade-filled regimens for coffee and cocoa cultivation because they would produce gains over sun-filled regimens in environmental services such as carbon sequestration.[31] Recent state initiatives in Ghana and Ivory Coast have made large numbers of new cocoa plants available to smallholders in the cocoa belts.[32] These initiatives could include the reestablishment of tree canopies in the renovated landholdings.

Ecocertification initiatives have produced a somewhat similar set of land use trends among smallholder coffee cultivators in Colombia.[33] A sizable minority of smallholder coffee growers opted for certified organic regimens. The certified farmers planted more trees and grew a wider array of tree species on their smallholdings than did comparable groups of smallholders who did not choose to be ecocertified. Here again, modest gains in environmental services, including carbon sequestration, occurred. A meta-analysis of mostly coffee ecocertification programs indicates that this pattern of environmental gains extends to other coffee-producing regions outside of Colombia.[34] In sum, the ecological devastation produced by rapid deforestation in these agroforestry belts has created a set of ecological conditions in which a modest recovery of environmental services would accompany the reestablishment of shade-filled regimens of agroforestry. Here, too, painstaking calculations of recovering environmental services might enable REDD+ like payments for increments in environmental services.

Dry Domestic Forests and Woodlands Among Impoverished Peoples

Very poor people in sub-Saharan Africa have, in circumstances where tree tenure is secure, created in increments of one tree at a time, domestic forests close to their homes. The continued existence and expansion of these forests depends on the work of families, so these forests can be considered "domestic" forests. Families plant most of the trees, and in semiarid settings, they water the trees. These forests spread across the sparse, semiarid woodlands of the Sahel between 1980 and 2010. NGO—farmer initiatives in these settings led to forest reforms which secured tree tenure for farmers. She or he who plants the tree owns it. Here, too, the labor for planting comes from households, and the trees, some of which sprout unassisted, provide a range of goods such as fruits and construction materials for families. Governments, through the Great Green Wall initiative, have provided assistance to smallholders by subsidizing the acquisition of seedlings.[35] A similar regional initiative has begun in the semiarid Miombo woodlands south of the Congo River basin.

The magnitude and duration of the gains in environmental services and household economies from these government initiatives depends on the maintenance of political stability in the region and on the continuing capacity and commitment of smallholders to care for the trees after planting.[36] The poverty of the people doing the planting and tree maintenance warrants the creation of a payment for environmental services (PES) program for domestic forests. The place of domestic forests in global carbon budgets may be small, but in their contribution to a global climate stabilization campaign, the efforts of impoverished domestic foresters have symbolic value.

A similar argument can be made for the contributions of urban foresters. In urban and periurban settings, smallholders provided the impetus for the first large tree planting campaigns in East Africa through the Green Belt Movement.[37] Most recently, these campaigns have scaled up to the globe and expanded their goals to one trillion trees. The institutional profile of these state, NGO, and farmer-managed efforts resembles corporatism, a fusion of global, national, and local interests sustained by periodic compacts about goals and commitments.

Community Forests

The political and demographic histories of rural communities have shaped the management of nearby forests in a wide variety of settings. For hundreds of years, elders in South Asian villages shaped the use of nearby forests. Independence from colonial regimes often left the administration of nearby forests in the hands of locals.[38] Tumultuous political transformations, such as the Mexican Revolution, may have actually strengthened the hands of local authorities in managing forest growth and regrowth. The production from community forests has varied from region to region. In Mexico, communities usually own the forests. Village collectives manage the forests and produce a wide range of forest products.[39] In Tanzania, village authorities authorize the collection of relatively few products, usually just charcoal and timber.[40] These community forests have persistently shown modest rates of growth in their extent.[41] The evolving governance of Indigenous preserves in other regions such as the Amazon basin may move, over time, in the direction of governance by communities.

Corporatist Processes, Polycentrism, and Regional Reforestation Dynamics

Three of the eight forestation processes depicted in figure 8.1 occur on all three continental land masses. People in Africa, the Americas, and Asia–Europe have protected forests (1), increased the number of large scale forest plantations (4), and developed more agroforests (6). Community forests (8) have shown modest increases in scale in the Americas and South Asia.

Mountainous terrain has induced secondary forest growth through land abandonment in the Americas and Asia–Europe (2), Three other forestation processes seem inextricably tied to particular locales: smallholder plantations in East Asia (3), silvopastures in the Americas (5), and dry domestic forests in sub-Saharan Africa (7). The processes shared across all three continental land masses reflect, no doubt, the workings of a single global political economy. The distinct locales for silvopastures, dry domestic forests, and smallholder plantations reflect the particular historical human ecologies of Africa, the Americas, and Asia.

Table 8.2 describes the most prevalent reforestation strategies in eight large-scale regions with forests. Clearly, one strategy does not fit all regions! Forest restoration in one region would mean something quite different in another region. The variety from region to region in forestation strategies underscores the accuracy of Elinor Ostrom's observation about the polycentric nature of efforts to mitigate climate change.[42]

Four common threads run through the natural climate solutions reviewed here. The first concerns political processes, in particular the corporatist political processes that have characterized successful efforts to restore or expand forests. The second concerns the two paths to forest gains (see figure 2.1). One path, the preservation of intact, old growth forests, brings with it a much wider range of ecological benefits than the second path whose variants rehabilitate degraded landscapes. The third thread notes that most of the restored forests have important human uses. They produce goods that are important to humans such as foods from agroforests

TABLE 8.2 Regional Patterns of Reforestation

Rural regions	Forestation strategies
Intact forests: boreal and tropical	Protected areas and Indigenous reserves
Montane regions: Asia, Americas, Europe	Abandonment of agricultural lands
Densely settled agricultural regions in asia	Smallholder forest plantations
Large farm agribusinesses	Large-scale forest plantations
Pastoral Amazon	Silvopastures
Coffee, cocoa, and rubber agroforests	Shaded agroforests
Dry tropics	Homestead managed restoration
Community forests	Community managed restoration

Sources: Case studies in chapter; see notes.

and wood products from plantation forests. The usefulness of these restored forests makes them easier to establish. The last of these analytic threads notes that most expanded or restored forests entail some sustainable agricultural intensification. Silvopastures increase the wood products from a tract of land without reducing the production of livestock from the land. Agroforests do something similar. Forest plantation monocrops increase the volume and frequency of harvests from degraded tracts of land. These intensified agricultural routines come at some ecological cost, but they seem unavoidable in a world with a growing human population.

While the comparative analyses of resurgent forests provide empirical support for the argument about the pivotal role of corporatist processes in preparing the ground for forestation, the same analysis underscores how difficult it can be to foster forest conservation, restoration, and expansion. Corporatist processes work the best in polities with established traditions of good governance. The double movement may play an important role here. Growing popular support around the globe for natural climate solutions would strengthen the government institutions that enable corporatist political processes. The next and final chapter explores how a global social movement to combat climate change might strengthen corporatist processes of governance, spur a global forest transition, and in doing so augment the role of natural climate solutions in stabilizing the climate.

• • • •

A Global Forest Transition?

If the Great Acceleration in human activity accounts for most of the losses of forests and biodiversity since the end of World War II, then a reverse trend, economic and demographic degrowth, might provide a path to forest recovery. Degrowth would trigger declines in demand for forest products. It would entail "an equitable down-scaling of production and consumption that [would] enhance environmental conditions."[1] Population decline in the form of below replacement fertility, another form of degrowth, promises an additional, relatively painless, albeit slow, way to reduce greenhouse gas emissions. This pattern of change would accelerate if cultural shifts toward voluntary simplicity in consumption patterns occurred along with fertility declines, as ethnographic studies suggest that they might.[2] Increasing numbers of conservationists recognize the link between personal consumption and the preservation of old growth forests. They ask like-minded people to limit their personal consumption in order to preserve the forests.[3] For voluntary reductions in consumption to curb consumer demand enough to diminish the amount of land in agriculture and induce reforestation, large segments of the middle classes in the Global South and the Global North would have to embrace restrictions on personal consumption.

A different, but not incompatible path to forest recovery would feature a global forest transition in which collective action by humans reverses the long decline in the global extent of forests. In scale, these transitions have, historically, characterized regions within nations or entire nations such as France, Denmark, or Switzerland.[4] In time, forest transitions have unfolded over periods as short as two decades and as long as a half-century.

When transitions occur in forests throughout the globe, then a global forest transition will have occurred. Arguments for natural climate solutions all presume that a global forest transition will take place. Otherwise, the gains in carbon sequestration from resurgent forests would be too small to make a significant contribution to climate stabilization. The current contrast between a northern hemisphere with increasing forests and a southern hemisphere with diminishing forests indicates that a global forest transition has not occurred. It is a normative formulation, a wished- for event contained in plans for landscape transformations but largely unrealized in actual changes in landscapes.[5]

If a global forest transition does occur, it will most likely take the form of a latecomer transition. Early and late transitions differ in the degree of planning entailed in them. The first forest transitions in the modern era (post-1800) usually occurred in an unselfconscious way when farmers, in response to a loss of labor or adverse trends in the prices of agricultural commodities, stopped cultivating some tracts of land. The fields, which farmers had cleared of trees during an earlier generation, then reverted to forests. Departures from this market-driven pattern did occur in several instances during the nineteenth century after heavy rains in deforested uplands caused downstream flooding in cities. In the aftermath of the floods, governments created subsidies to spur reforestation in the uplands.[6] In general, though, the early forest transitions occurred slowly over relatively small areas, without explicit plans to convert pastures or croplands into forests.

Late transitions, of which an emerging global forest transition would be a prime example, take a different form. Alexander Gerschenkron, an economic historian, outlined the dynamics of late transitions.[7] Unlike early transitions, late transitions have an audience of analysts who assess the desirability of the observed change. If they consider it desirable, the analysts promote government policies to expedite the change. Gerschenkron described how this dynamic played out during the industrialization of Western Europe in the nineteenth and early twentieth centuries. Great

Britain experienced an early transition to an industrialized economy without a set of government policies designed to expedite the mass production of manufactured goods. European competitors of the English quickly appreciated the advantages of mass production. In essays and speeches, European leaders emphasized the need to catch up or keep up with the British. To this end, leaders tried to replicate the British experience with a set of government policies. The Germans, after unification, proved particularly diligent in this regard. Through centrally directed policies, they achieved very rapid rates of industrialization during the three decades before World War I. Their late transition differed from the early English transition in three ways: (1) The Germans appreciated the need to industrialize, so they planned to do so. (2) The centralized power of the state played an important role in implementing the plan. (3) The Germans industrialized more rapidly than did the British.

Recent, national-level forest transitions have exhibited the characteristics of late transitions. Crisis narratives drove the transitions. They mobilized the policymakers who then created plans to address the crises through state actions. In China, for example, the devastating floods along the Yangtze River in 1998 provided the political impetus behind the "grain for green" reforestation program. In Vietnam shortages of wood in the urban and industrial sectors of the economy generated commitments from government officials to increase forest cover. In both instances, the central governments embarked on premediated, aggressive expansions of forests that insured rapid change. Most of the increases in forest cover in China and Vietnam came in the form of state-supported expansions in forest plantations.[8] Change occurred rapidly. In a fifteen- to twenty-year period, government programs reversed long established deforestation trends.

While differing trends in forest cover between the Global North and South indicate that a global forest transition has yet to occur, the preconditions for a global, latecomer forest transition do exist. In the first decades of the twenty-first century, the crisis narratives surrounding the need for natural climate solutions grew more compelling and, in response, government officials set out ambitious plans to reduce greenhouse gas emissions. Beginning with the 2015 IPCC Conference in Paris, governments have submitted so-called nationally determined contributions (NDC) plans for reducing their emissions.

Increases in forest cover figured prominently in the NDC plans of many countries. Of the 158 countries that submitted plans, 71 (44.9 percent) relied on carbon sequestration in expanding forests to cover a portion of their overall pledged reductions in greenhouse gas emissions. In some countries, local experiences with forest resurgence shaped their plans for future forest expansion. Niger and Nepal both opted for expansions of existing, successful forest restoration initiatives led by local communities and NGOs. In other countries such as Belize and Papua New Guinea, officials decided to create national REDD+ programs after participating in extensive discussions of REDD+ at conferences.

The projected forest transitions in these plans would occur through centralized political processes that feature rapid, government-funded expansions of forest plantations. The political imperative of making rapid progress probably accounts for the popularity of forest plantations as a means for forest expansion among the countries that joined the Bonn Challenge for Forest Restoration.[9] The Paris NDC plans, considered together, anticipate a global forest transition. It has the recognizable features of a latecomer transition: a crisis-driven plan, with centralized implementation, and the promise of a rapid rate of change. The impetus behind these plans for a global transition comes in large part from the millions of people now mobilized into a transnational social movement to limit climate change.

The Transnational Climate Movement: New Norms, Direct Action, and Behavioral Changes

I . . . have a dream.
That governments, political parties, and corporations
grasp the urgency of the climate and ecological crisis
and come together despite their differences—as you would
in an emergency—and take the measures required to safeguard
the conditions for a dignified life for everybody on earth.

—GRETA THUNBERG

At the beginning of the 1980s, a flurry of publications by ecologists in pres-
tigious outlets voiced alarm over the rapid destruction of tropical rainforests.[10] These books and articles marked the beginnings of a countermovement dedicated to reversing the environmental destruction of forests from
the Great Acceleration in human activities after World War II. During the
next few years, the alarmed voices expanded into a chorus, and a transnational social movement emerged with the goal of preserving tropical rainforests. The movement mobilized scientists in universities, activists in new
NGOs, and philanthropists in foundations to pressure for the preservation
of old growth forests. Movement leaders raised funds to purchase forested
land. The purchased land then became protected areas. By the end of the
1980s, movement activists had begun to make common cause with Indigenous leaders in efforts to establish forested preserves for Indigenous peoples. By the second decade of the twenty-first century, the movement's
momentum had begun to dissipate. The expansion in protected areas had
slowed down in the Americas.[11] Influential observers lamented, as noted
earlier, that the 2010s represented a lost decade in efforts to preserve forests.[12] Still, in this uncertain moment for the movement, it has the potential, through the application of political pressure, to trigger a global forest
transition.

A new, global coalition of activists has emerged. Activists in the larger
movement to limit climate change have joined forces with activists in
the movement to protect rainforests. The merger of climate change and
biodiversity activists has been most evident during the annual Conference
of Parties (COP) meetings. The assembled nations of the 2021 Glasgow COP
pledged to end deforestation by 2030 in what became the most newswor-
thy event of the meeting. Countries and companies committed a combined
18 billion dollars to fund this effort. The fusion of climate change and biodiversity concerns also surfaced in the UN 2020 Intergovernmental Science-
Policy Platform on Biodiversity and Ecosystem Services.[13] While faction-
alism may eventually rob this coalition of its most dynamic elements, it
has not had debilitating effects to date.

Activists in the movement have argued for changes in norms among
consumers, producers, and politicians that would slow and maybe even
reverse the destruction of forests. Normative changes, as embodied in the

voluntary simplicity movement, could diminish demand for forest products and reduce forest losses. Potentially important normative changes are visible in diets. Declines in the consumption of beef could spur the abandonment and reforestation of old cattle pastures. Beef consumption in the United States has experienced a significant decline since 1970. It decreased from 84 pounds per capita in 1970 to 58 pounds per capita in 2020. If this trend becomes more widespread, as suggested by surveys that show increased self-identification as vegetarians among Indian, Chinese, and American consumers, then cattle ranchers, faced with declining prices for beef, might allow some of their pastures to revert to forests, as occurred in Costa Rica during the 1990s.[14]

The recent promotions of non-beef burgers at fast-food establishments in the United States stems most likely from a concern with the environmental effects of converting extensive amounts of forests into pastures for cattle. This change in menus resulted from the persistent lobbying of fast food executives by rainforest activists who argued for the accuracy of a "hamburger thesis" that eating hamburgers destroys rainforests.[15] Changing a menu does not guarantee the massive change in diets that would lead to the large-scale reversion of tropical pastures into forests, but it does represent an initial step in that direction.

Other consequential changes in norms have occurred in the supply chain for beef. Changes have occurred in the confinement of cattle, with consequences for forest clearing. Confined cattle consume less forage than grazing cattle, and the additional feed consumed by confined cattle comes from relatively small areas of cropland, so the overall need for agricultural land may decline when people confine cattle. Environmentalists made this case forcefully after 2000, and some Brazilian cattle ranchers in the southern Amazon basin responded to these pressures by constructing more confinement facilities.[16] Comparisons of farms with and without confinement facilities indicate lower deforestation rates on landholdings with confinement facilities than on landholdings without them. Landholdings with confinement facilities also had more cropland than comparable landholdings without confinement facilities. The influence of the confinement facilities on land uses extended beyond the boundaries of landholdings with facilities. Properties close to confinements contained more cropland for grains such as soy and deforested less land for cattle pastures than did properties far from confinement facilities.[17] In this instance, movement activists have contributed to a shift in norms about raising cattle in confinement that appear to have reduced cattle-related greenhouse gas emissions by conserving more forest on ranches.

The spread of forest enhancing norms has also occurred through growth in the extent of forests with long-term management plans. As the extent of forests included in protected areas, Indigenous preserves, and forest concessions has increased, long-term forest management has become more common. By 2020, long-term management plans covered 54 percent of the globe's forest area.[18] The increase in managed forests underscores several trends in the human–forest interface. First, the spread of norms about forest management indicates a growing acceptance of the importance of issues of sustainability in the use of forests. Second, the spread of managed forests signals a decline in "open access" forests that permit unrestricted use, including deforestation. To be sure, open access forests continue to exist. Approximately one-third of the Brazilian Amazon remains accessible to purchasers, and almost all of the recent destruction of Amazon forests has occurred in these unrestricted areas. Nonetheless, the expansion in managed forests makes it more difficult for loggers, planters, and farmers to find forests for clearing. Under these circumstances, more groups may have resorted to illegal logging, especially in societies with weak governance.

Construed more generally, a "world society," with norms that would increase the likelihood of a forest transition, has begun to emerge.[19] This dynamic seems apparent in the consultations and revisions that accompanied the submission of NDC plans at the 2015 Paris COP meeting. Neighboring nations submitted similar plans, so regional clusters of planned forest expansion in, for example, the Atlas Mountains of North Africa or montane districts in East Africa became explicable as "duplicates of what the neighbors are doing."[20] Pan regional norms about land use also spread in settings far from global meetings. For example, as noted in chapter 6, governments in the Sahel organized "the Great Green Wall" in 2008 in an attempt to accelerate the spread of woodlands in the semiarid region just to the south of the Sahara Desert. Ten years later, governments in the semiarid regions to the south of the Congo River basin organized a second Great Green Wall initiative to stem the advance of the Kalahari Desert into districts with dry forests.[21]

The spread of norms and practices, especially if propelled by a social movement, often seems to occur in waves. As noted earlier, the programs to facilitate the creation of forest plantations have come in waves. China launched its massive "grain for green" program to plant forests in the upper reaches of the Yangtze River watershed after the devastating floods of 1998. The Vietnamese, having drawn their supplies of wood down to low levels after decades of cutting during the late twentieth century, decided to follow the Chinese example. A wave of tree planting swept across Southeast Asia. The adoption of ecocertification schemes in agroforestry, noted in table 7.3, occurred in a wavelike tempo between 2010 and 2016, leading to a rapid expansion in the flow of ecocertified goods. Social movement activists accelerate these changes through persuasion and alliances with like-minded politicians.

The combination of normative change driven by crisis narratives and policy experiments leave open the possibility that, together, they could induce a global forest transition. It is, however, hard to imagine this kind of scaling up in forest conservation, restoration, and creation without the impetus provided by a vigorous transnational social movement dedicated to limiting warming to 1.5°C. In other words, a global forest transition seems unlikely to happen in the absence of a transformative social movement dedicated to curbing climate change. Activists would further climate solutions by advocating for normative changes and by institutionalizing landscape changes through corporatist processes.

Corporatism: Institutionalizing a Social Movement for Forests

Over time, the actions of the countermovement, like those outlined earlier, have created corporatist infrastructures that promote the resurgence of forests. Let us consider five examples of how activists in the movement have created, in agitating for environmental protection, corporatist-like political processes that have fostered forest growth and regrowth. Examples of this dynamic come from a wide range of social scales: local coalitions that contest with real estate developers about proposals to clear relatively small parcels of forested land, Indigenous alliances that contend with extractive industries across a region, national programs that promise to strengthen REDD+ programs that pay for environmental services, ecocertification schemes that over time have become more environmentally restrictive, and the global meetings of the IPCC's Conference of Parties that, despite formidable opposition, has gotten nation states to acknowledge new climate stabilization responsibilities.

Land Use Controversies and Conservation Coalitions

In the examples cited in chapter 3, local opponents of rural real estate development contended successfully with wealthy economic interests only because, repeatedly, they secured assistance from governments and well-financed, oftentimes international, environmental organizations. These types of links between individual landowners, local governments, and global organizations characterize societal corporatist processes. The mobilization of opposition to rural and periurban real estate development through these links lays the political groundwork for the creation of new protected areas. After a period of conflict, the antagonists settle on a deal. At the cost of some new development, conservation activists get a new protected area, albeit smaller in size than the area that they initially mobilized to defend. The conflict leaves a legacy in the form of social bonds between coalition members that facilitate the formation of another coalition to negotiate an agreement around the next proposal for real estate development in a place. Easy to mobilize, these quasi-corporatist regional conservation partnerships have formed in places with long histories of conflict over land uses.

The Global Indigenous Movement

A similar trajectory of global institutional development has characterized Indigenous groups. Beginning in the 1960s, Indigenous communities created organizations to represent their interests and to secure titles to forested land. Over time, activists built alliances across these Indigenous organizations. They now make common cause with each other and with like-minded conservationists. To this end, fifteen Indigenous groups joined the two hundred signatories of the 2014 New York Declaration on Forests that pledged to end forest losses by the 2020s.[22] Most recently, activists have created global alliances of Indigenous peoples who have pressed for more forest protection. For example, representatives of the Global Alliance of Territorial Groups pressured delegates from the wealthier nations to do more for forests at the Glasgow COP in 2021.

Aggregate data underlines the significance of the Indigenous sector for the conservation of global forests. As of 2020, Indigenous peoples controlled about 18 percent of the world's forest area. Approximately 1.5 billion Indigenous and local peoples have acquired collective property rights over forested land since the 1970s. Indigenous dominion over land has concentrated in regions with intact forests. Indigenous peoples currently serve as custodians for nearly 40 percent of the ecologically protected, intact forests on earth.[23]

Indigenous groups have frequently taken direct actions to press their arguments for the adoption of forest-friendly norms. Episodes of unruly behavior and a willingness to take direct action have spurred change in many instances.[24] For example, in 1988 leaders from Brazil, along with an American anthropologist, Darrell Posey, traveled to Washington, DC, to protest loans given by the World Bank to Brazil to finance dam construction on the Xingu River. Their testimony delayed the loans. Protests by other Indigenous leaders that same year led the World Bank to withdraw its financial support for road paving projects in the southwestern Brazilian Amazon.[25] Similarly, the Shuar in the Ecuadorian Amazon have shown a willingness to confront mining interests. In 2006 a group of Shuar families occupied a mining camp and shut it down. Their actions stopped the development of a copper mine for ten years, until 2016 when the Ecuadorian Army ousted the Shuar forcibly from the mining camp.[26]

Global coalitions of Indigenous activists have increased the probability that even small groups of Indigenous peoples, contending with the threat of forest-destroying outside investors, can assemble coalitions of like-minded activists to defend their forests. In addition, these organizational efforts have made it easier at a base level to create communities of "forest stewards" who, in defending their livelihoods, preserve the forests in which they reside.[27] As in other instances of conflict over forest conservation, a range of Indigenous representatives from large-scale and small-scale organizations have come together to promote forest preservation and recovery.

REDD+ Projects

Growth in the strength of the climate stabilization movement would, through a series of political and technological developments, strengthen REDD+ efforts at forest conservation. A lack of demand from greenhouse gas emitters in wealthier countries reduced the price for a ton of sequestered carbon to as low as three to five dollars a ton during the 2005–2020 period. The low prices in turn reduced the funds that REDD+ administrators had on hand to pay landowners for the carbon sequestration on their lands. Without a robust flow of funds from the purchasers of offsets or a clearly identified set of landowner-recipients for REDD+ funds, many administrators opted to take a "jurisdictional approach" to REDD+.[28] These jurisdictional programs promoted sustainable land use management across government jurisdictions, allowing, for example, for the public financing of tree nurseries or subsidies to facilitate the purchase of more productive forages for livestock. This widening of the array of movement supported projects and actions represents a politically feasible, albeit less efficient way to advance the cause of rainforest protection. These projects resemble the much discussed and critiqued "conservation and development" projects of the 1990s.[29]

Some of these obstacles to the implementation of REDD+ projects may disappear in the near future. A relatively large number of development institutions, including the World Bank and bilateral assistance programs from Western Europe, have prioritized programs to reduce land tenure insecurity.[30] These programs have been large enough in scope to address the tenure insecurity impediment to implementing REDD+ in forest-rich nations such as Peru and Indonesia. Funds for carbon offsets may also increase. Some analysts project a tenfold increase in the price of carbon offsets by 2030, as more companies make zero-emission pledges.

Improved surveillance through remote sensing should make it easier to protect forests and reward conservation minded landowners through REDD+ programs. It becomes possible, as Brazilian regulators have demonstrated, to identify even small, recently deforested tracts of land. This capacity in turn enables the enforcement of laws against the destruction of forests. The high-resolution imagery could also be of use in decommissioning logging roads once the roadside lands have been logged. Destroying these roads would prevent the later conversion of the cutover lands into agricultural fields.

Accurate maps of forest–farm mosaics may also spur durable improvements in the security of land tenure, which in turn would encourage forest preservation through REDD+ projects. Satellite-derived maps, given their accuracy and detail, encourage the clarification of questions about land use and ownership. Leaders began Indonesia's One Map initiative with this kind of clarification in mind.[31] High-resolution satellite imagery of disputed areas would encourage a corresponding clarification of landowner interests and responsibilities, which in turn would contribute to longer term commitments to tracts of land by occupiers, with concomitant adoption of more sustainable land uses.

Further efforts to institutionalize REDD+ at an unprecedented scale occurred at the 2022 Sharm el Sheikh COP of the IPCC. The three countries with the most intact tropical rainforests, Brazil, the Congo, and Indonesia, agreed to create a REDD+ like funding mechanism to preserve their tropical rainforests. It increases the scale of the REDD+ jurisdictional approach to a transnational level. The three national governments would funnel REDD+ payments to projects intended to preserve forests within their borders. Slippage in these transfer payments could occur given the continuing vitality of the forest destroying, business-as-usual growth coalitions in the three countries. Still, this initiative does represent another instance of transnational mobilization, one that promises to increase the salience and scope of the REDD+ mechanism for preserving forests.[32]

The clarity of the REDD+ system of payments makes it intelligible and applicable across a wide variety of contexts. Its implementation depends on the mobilization of the larger climate change movement and its ability to influence corporate actors who purchase carbon offsets and government officials who secure land tenure for smallholders in forest-rich countries. A corporatist-like coalition of international donors, remote sensing specialists, land tenure officials, and local landowners would facilitate the implementation of REDD+ projects. In effect, the REDD+ system would become institutionalized under these conditions.

Ecocertification Organizations

Ecocertification schemes offer another management tool, organized in corporatist processes, that in scaled-up form could induce a considerable increase in the volume and variety of environmental services provided by agroforests. Because consumer demand for the most economically important agroforestry products such as coffee, cocoa, and rubber is elastic, intensified production does not necessarily produce land sparing in agroforestry districts.[33] Environmental recovery in these contexts would have to occur through the spread of norms about cultivation that minimize disturbances in agroforestry landscapes. Ecocertifying organizations such as the Rain Forest Alliance or the Forest Stewardship Council provide a vehicle for disseminating these environmentally friendly norms.

Ecocertification efforts have had a mixed historical record in the first two decades of the twenty-first century. Among the cases discussed here, the FSC failed to prevent the construction of roads in logging concessions in the DRC, but it did produce environmental gains in Chile's logging concessions. The early history of the Roundtable for Sustainable Palm Oil (RSPO) suggests a mix of conclusions. Because RSPO certified plantations contained so few tracts of primary rainforest when they became ecocertified, they had little impact on the overall rate of deforestation in oil palm producing districts.[34] Insistence on maintaining deforestation-free histories has, however, over time made RSPO certification more meaningful. To earn bragging rights with consumers, growers would have to maintain a

clean bill of health with the RSPO over a continuous period of time, beginning in 2005. It also becomes possible in this context to imagine a more
environmentally consequential set of ecocertifications in which certifiers
induce growers to rehabilitate degraded lands such as grasslands into sustainably produced oil palm groves or cacao agroforests.

Ecocertification appears to have spread among growers in geographically
uneven ways, common in the Americas and more unusual in South and
Southeast Asia. Colombia contains a large concentration of certified coffee
producers, while Ecuador grows three-quarters of the world's certified
bananas. These regional disparities in certifications may reflect regional differences in consumer preferences for certified goods. While growers in the
Americas may produce for wealthy, environmentally concerned consumers in Europe and the Americas, growers in South and Southeast Asia produce largely for poorer Asian consumers who have less interest in consum-
ing expensive, sustainably produced goods. A world societal trend in which
more consumers opt to purchase ecocertified goods would magnify the
impact of ecocertification programs.

IPCC Conferences of Parties

Finally, the annual meetings of the IPCC's Conference of Parties show signs
of evolving into a corporatist process. Since 2008, the expectations of participants in the meetings have risen. In the first meetings, only the wealthier countries appeared willing to make commitments to reduce emissions.
All participants in the 2009 Copenhagen meetings agreed to reduce emissions in accordance with their individual situations. Countries at the 2015
Paris COP agreed to submit plans for making further emission reductions
every five years. Most recently, country participants in the 2021 meeting
in Glasgow agreed to submit plans for additional emission reductions at
every annual COP meeting.

While country delegations did not make binding commitments at any
of the COPs, the meetings' leaders did ask for more from the signatories
over time. In effect, the agreements at the COPs appear to be moving toward
annual compacts in which different participants, companies as well as

states, make commitments that they intend to honor. In this respect, the COPs, in their participants, bargaining sessions, expectations, and agreements, have begun to resemble a corporatist process. Like the local land use controversies, the Indigenous alliances, the REDD+ projects, and the ecocertification processes, the COPs have become more institutionalized in their expectations of participants. The pressure to participate in these compacts comes from the global climate movement.

Polycentric Governance

The looming catastrophe of uncontrolled climate change recalls other difficult circumstances that have called for the mass mobilization of societies. One occurred in the immediate aftermath of World War II when Japanese households, with numerous mouths to feed, tried to patch together livelihoods without the means to do so. In this context, Kingsley Davis, an American demographer, described, with evident admiration, the thoroughgoing efforts of the Japanese people to reduce their fertility rate during the first decade after their defeat. As Davis put it, the Japanese people "responded in every manner then known to some powerful stimulus."[35] They delayed marriage, got abortions, and sponsored family planning services, all in an attempt to bring down fertility rates and reduce household poverty. They left no path unexplored in their efforts to reduce fertility. A similar polycentric approach should characterize efforts to increase the magnitude of the forest carbon sinks in the twenty-first century.[36]

David Kaimowitz's point about the gains to be had in the efficacy of political pressure becomes pertinent here.[37] When initiatives build on local trends that are already under way, they have a greater impact than they would have had otherwise. Following this dictum, advocates for forests would engage with local landholders in forest regrowth initiatives that are already under way. Corporatist political processes would enable these regionally distinct links between global and local interests.

The leaders of the IPCC effort to minimize climate change institutionalized this polycentric approach to climate governance in 2015 when they called for each country to create and submit a plan for Nationally

Determined Contributions (NDCs) to a global mitigation effort. The planners in each country assessed how, given their natural resource endowments, they could cut their greenhouse gas emissions and/or sequester sufficient amounts of carbon to achieve carbon neutrality by a specified date (usually 2050 or 2060) in the future. Because resource endowments such as forests, winds, and flowing water vary regionally, with important implications for generating energy and sequestering carbon, plans for reducing emissions from energy expenditures, outlined in the NDCs, should vary regionally. Accordingly, escalated efforts to sequester more carbon through forest restorations would vary with the type and extent of forest cover across regions. National climate solutions would accordingly take a polycentric form.

Corporatism, Income Redistribution, and a Global Forest Transition

In its architecture, a global forest transition would resemble a latecomer transition. A crisis narrative would drive the process. International elites would set a broad agenda. National governments would plan the change. It would, according to the plans, occur quickly. A corporatist political process would lead to more effective implementation of the plans. Local governments would implement the plans, working in concert with crucial occupational groups. Farmers, both large and small, would have to play an important role in these landscape changes. They have cleared the largest amounts of tropical forest during the twentieth and twenty-first centuries, so they should play a central role in global efforts to restore forests in the mid-twenty-first century. Through discussions, observations, and repetitions of time-honored routines, farmers have shaped the norms that govern land uses. In many instances, having just cleared forests from the land, farmers either oppose or are indifferent to forest restoration efforts. In contrast, high-profile assemblages of environmentalists, scientists, and government officials have argued vehemently for programs to forestall further deforestation and restore forests.

To expedite a global forest transition, activists would need to build robust organizational links between these two groups. The mandate for these links would come from agreements among nations, building on the Paris 2015 compact, to mitigate climate change through reduced greenhouse gas emissions. A massive societal infrastructure would have to emerge across nations, a kind of societal corporatism. Participants in these compacts would become parties to a new social contract that would bind participants in these corporatist processes to commitments that they make in negotiating the compacts.[38]

These agreements would tie farmers and farming organizations with norms about land use to global organizations with goals of reducing greenhouse gas emissions through forest conservation, restoration, and expansion. By bringing together these disparate elements and negotiating compacts between them, corporatist processes offer an optimal organizational framework for achieving natural climate solutions. The specific organizational arrangements vary across different arenas for collective action, from local land use controversies to REDD+ projects and IPCC meetings. Three elements characterize all of these corporatist arrangements. They bring together disparate social groups; the groups negotiate compacts that commit the different parties to climate friendly collective actions, and the groups recreate these compacts repeatedly over the years. As outlined in earlier chapters, many of the compacts engineer or encourage, in one way or another, the reforestation of degraded lands. The political feasibility of these organizational transformations may depend on two imponderable developments.

First, will the pressures for degrowth and the spread of a voluntary simplicity ethic among consumers, described briefly at the outset of this chapter, become significant enough to reduce the economic pressures for agricultural expansion? A reduction in the opportunity costs associated with forest conservation and restoration would follow from the reduced human pressures on natural resources. In this context, farmers might find a subsidized reversion of fields into forests more palatable.

Second, would any redistribution of income occur with the implementation of forest enhancing programs? Payments for environmental services (PES) have figured centrally in discussions of forest conservation, restoration, and expansion efforts. The distribution of these payments to landowners could follow a REDD+ model in which payments to landowners occur in rough proportion to the volume of carbon sequestered on the owner's land during the preceding year. With these kinds of rules for payments, forest landholdings concentrated in a few hands would produce very unequal REDD+ payments. An unjust forest transition would occur.

In some instances, concentrated PES–REDD+ payments would serve social justice ends. Impoverished Indigenous peoples with extensive, forested landholdings would receive relatively large PES payments from donors with concentrated wealth. In many other circumstances, wealthy private and corporate entities own large amounts of forested land, and they would presumably receive the largest PES payments. The association between wealthy interests and conservation has a long history. Benefactors have purchased forested lands and then donated them to governments which created protected areas out of the donated lands. In some instances, a pattern of coercive conservation emerged in which the state or wealthy private landowners have evicted poor people from newly established protected areas.[39]

This last distributional pattern does not bode well for the cause of forest conservation and restoration because it provides only small, tangible, short-term benefits, aside from climate stabilization, to most rural peoples. Under these circumstances, the 2017 yellow vest protests in France, in which rural peoples opposed climate change restrictions, become easy to understand. Amplified payments to smallholders offer one way out of this political impasse. PES–REDD+ projects would be negotiated within a corporatist framework and subject to revisions from year to year depending on the record of carbon sequestration during the preceding year. These arrangements would conceivably redistribute income from wealthy consumers in the Global North to poor producers in the Global South. A just global forest transition would occur.

The political calculus of support for natural climate solution programs would change accordingly. Support for forest protection would presumably grow in the Global South. While these measures would reinforce a global forest transition, they would do little or nothing about the resistance from fossil fuel elites that, to date, have prevented widespread reductions in greenhouse gas emissions. While reducing greenhouse gas emissions from fossil fuel–powered machines remains the primary political-economic challenge facing the climate stabilization movement, just natural climate solutions, through negative emissions from forest regrowth, can make an important contribution to our emerging effort to stabilize the earth's climate.

Glossary

AFFORESTATION: Establishment of forests by planting trees on lands historically without trees.

AGROFORESTS: The cultivation of row crops, shrubs, and trees on a single tract of land.

CORPORATISM: Governance through decision making by organizations that include all participants in a sector. In the landscapes sector, politicians, merchants, NGO activists, donors, farmers, foresters, landholders, and landowners make periodic decisions to concert their actions.

DEFORESTATION: Land cleared of all trees for a long period. Land cleared of trees by loggers that then regenerates into forest would not be an example of deforestation.

FOREST: Land with tree crowns that cover more than 10 percent of the land area.

FORESTATION: Net forest cover gains that exceed net forest cover losses over time in a place.

FOREST PLANTATIONS: A stand of planted trees, for harvest and sale as soon as commercially feasible.

FOREST TRANSITIONS: Net forest cover gains replace net forest cover losses over time in a place.

GLOBAL FOREST TRANSITION: A global-scale forest transition.

LATECOMER TRANSITIONS: Changes that occur after similar changes in other places. The later changes tend to be more planned, more centrally directed, and more rapid.

NATURAL CLIMATE SOLUTIONS: Conservation, restoration, and land use conversion that increase carbon storage and in doing so contribute to declines in greenhouse gas emissions that will end the climate crisis.

NEGATIVE EMISSIONS: Carbon dioxide removal from the atmosphere, in this instance through the sequestration of carbon in the biomass of forests.

REFORESTATION: Agricultural land where, after many years, trees regenerate, forming tree cover canopies in excess of 10 percent of the land. When trees regenerate naturally in a place, forest restoration and reforestation are equivalent. At a global scale, such as "reforesting the earth," reforestation refers to an increase in tree cover on the earth's land area. At large scales, the terms "forestation" and "reforestation" can be used interchangeably.

SECONDARY FORESTS: Trees that regrow spontaneously on former agricultural land.

SILVOPASTURES: Pastures with significant numbers of trees.

Notes

1. Forests

1. B. Griscom et al., "Natural Climate Solutions," *Proceedings of the National Academy of Sciences* 144 (2017): 11645–50; Bernardo Strassburg et al., "Global Priority Areas for Ecosystem Restoration," *Nature* 586 (2020): 724–29.
2. National Academies of Sciences, Engineering, and Medicine, *Negative Emissions Technologies and Reliable Sequestration: A Research Agenda* (Washington, DC: The National Academies Press, 2019).
3. Griscom, "Natural Climate Solutions."
4. X. Song et al., "Global Land Change from 1982 to 2016," *Nature* 560 (2018): 639–44.
5. C. Chen et al., "China and India Lead in Greening of the World Through Land-Use Management," *Nature Sustainability* 2 (2019): 122–29.
6. N. McDowell et al., "Pervasive Shifts in Forest Dynamics in a Changing World," *Science* 368, no. 9463 (2020): eaaz9463.
7. Aiguo Dai, "Increasing Drought Under Global Warming in Observations and Models," *Nature Climate Change* 3 (2013): 52–58.
8. Aiguo Dai, "Drought Under Global Warming: A Review," *WIRES—Climate Change* 2 (2011): 45.
9. C. Allen et al., "A Global Overview of Drought and Heat-Induced Tree Mortality Reveals Emerging Climate Change Risks for Forests," *Forest Ecology and Management* 259, no. 4 (2010): 660–84.
10. L. Tacconi, *Fires in Indonesia: Causes, Costs, and Policy Implications* (Bogor, Indonesia: Center for International Forestry Research, 2003).
11. Tiago Oliveira et al., "Is Portugal's Forest Transition Going Up in Smoke?," *Land Use Policy* 66 (2017): 214–26.
12. Andrei Khalip, "Portuguese Protest Over Deadly Forest Fires: Government Pledges Aid," Reuters, October 21, 2017, https://www.reuters.com/article/us-portugal

-fire-government/portuguese-protest-over-deadly-forest-fires-government
-pledges-aid.

13. Eric Post and Michelle Mack, "Arctic Wildfires at a Warming Threshold," *Science* 378, no. 6619 (2022): 470; Adria Descals et al., "Unprecedented Fire Activity Above the Arctic Circle Linked to Rising Temperatures," *Science* 378, no. 6619 (2022): 532–37.

14. N. Andela et al., "A Human-Driven Decline in Global Burned Area," *Science* 356 (2017): 1356–62.

15. Frances Seymour and Jonas Busch, *Why Forests? Why Now? The Science, Economics, and Politics of Tropical Forests and Climate Change* (Washington, DC: Center for Global Development, 2016); John W. Reid and Thomas Lovejoy, *Ever Green: Saving Big Forests to Save the Planet* (New York: Norton, 2022).

16. Strassburg et al., "Ecosystem Restoration."

17. Griscom et al., "Natural Climate Solutions."

18. Thilde Bruun et al., "Environmental Consequences of the Demise in Swidden Cultivation in Southeast Asia: Carbon Storage and Soil Quality," *Human Ecology* 37 (2009): 375–88.

19. S. Cook-Patton et al., "Mapping Carbon Accumulation Potential from Global Natural Forest Regrowth," *Nature* 585 (2020): 545–50.

20. B. Bernal, L. Murray, and T. Pearson, "Carbon Dioxide Removal Rates from Forest Landscape Restoration Activities," *Carbon Balance Management* 13 (2018): 22.

21. Griscom et al., "Natural Climate Solutions."

22. R. Dave et al., *Second Bonn Challenge Progress Report: Application of the Barometer in 2018* (Gland, Switzerland: IUCN, 2019); NYDF Assessment Partners, *Protecting and Restoring Forests: A Story of Large Commitments yet Limited Progress* (New York: Climate Focus, 2019). Accessible online at forestdeclaration.org.

23. Dave et al., "Second Bonn Challenge."

24. R. Rajão et al., "The Rotten Apples of Brazil's Agribusiness: Brazil's Inability to Tackle Illegal Deforestation Puts the Future of Its Agribusiness at Risk," *Science* 369, no. 6501 (2020): 246–48.

25. C. Garcia et al., "The Global Forest Transition as a Human Affair," *One Earth* 2 (2020): 417.

26. Griscom et al., "Natural Climate Solutions."

27. Griscom et al., "Natural Climate Solutions."

28. Amarta Sen, "The Ends and Means of Sustainability," *Journal of Human Development and Capabilities* 14, no. 1 (2013): 6–20.

29. Holly Gibbs et al., "Tropical Forests Were the Primary Land Source for New Agricultural Lands in the 1980s and 1990s," *Proceedings of the National Academy of Science* 107, no. 38 (2010) 16732–37.

30. Thomas Rudel et al., "Changing Drivers of Deforestation and New Opportunities for Conservation," *Conservation Biology* 23 (2009): 1396–1405; Kemen Austin et al., "Trends in Size of Tropical Deforestation Events Signal Increasing Dominance of Industrial-Scale Drivers," *Environmental Research Letters* 5 (2017): 054009.

31. David B. Bray, *Mexico's Community Forest Enterprises: Success on the Commons and the Seeds of a Good Anthropocene* (Tucson: University of Arizona Press, 2020).

32. International Union for the Conservation of Nature (IUCN), *Boosting Biodiversity in Colombia's Cattle and Coffee* (Gland, Switzerland: IUCN, 2022).

33. David Kaimowitz, "The Meaning of Johannesburg," *Forest News*, CIFOR, September 19, 2002.

34. Karl Polanyi, *The Great Transformation* (New York: Farrar and Rinehart, 1944); John McNeill, *The Great Acceleration: An Environmental History of the Anthropocene Since 1945* (Cambridge, MA: Harvard University Press, 2016).

2. Theory

1. Food and Agricultural Organization (FAO) and United Nations Environmental Program (UNEP), *The State of the World's Forests 2020: Forests, Biodiversity and People* (Rome: Food and Agricultural Organization of the United Nations, 2020).

2. Food and Agricultural Organization, *Global Forest Resources Assessment, 2020* (Rome: Food and Agriculture Organization of the United Nations, 2020).

3. Peter Murphy and Ariel Lugo, "Ecology of Tropical Dry Forest," *Annual Review of Ecology and Systematics* 17 (1986): 67–88.

4. Precipitation records are taken from https://www.indexmundi.com/facts /indicators/AG.LND.PRCP.MM/rankings. See Robert Zomer et al., "Global Tree Cover and Biomass Carbon on Agricultural Land: The Contribution of Agroforestry to Global and National Carbon Budgets," *Science Reports* 6 (2016): 29987.

5. Food and Agricultural Organization, *Mountains as the Water Towers of the World: A Call for Action on the Sustainable Development Goals (SDGs)* (Rome: Food and Agricultural Organization of the World, 2014).

6. A. Sofia Nanni and H. Ricardo Grau, "Agricultural Adjustment, Population Dynamics and Forest Redistribution in a Subtropical Watershed of NW Argentina," *Regional Environmental Change* 14 (2014): 1641–49.

7. Michael Williams, *Deforesting the Earth: From Prehistory to Global Crisis* (Chicago: University of Chicago Press, 2006).

8. Robert Müller et al., "Spatiotemporal Modeling of the Expansion of Mechanized Agriculture in the Bolivian Lowland Forests," *Applied Geography* 31, no. 2 (2011): 631–40.

9. Y. Polain de Waroux et al., "Rents, Actors, and the Expansion of Commodity Frontiers in the Gran Chaco," *Annals of the American Association of Geographers* 108, no. 1 (2018): 204–25.

10. Karina Winkler et al., "Global Land Use Changes Are Four Times Greater Than Previously Estimated," *Nature Communications* 12 (2021): 2501.

11. Patrick Meyfroidt and Eric Lambin, "The Causes of the Reforestation in Vietnam," *Land Use Policy* 25, no. 2 (2008): 182–97.

12. Patrick Meyfroidt and Eric Lambin, "Global Forest Transition: Prospects for an End to Deforestation," *Annual Review of Environment and Resources* 36 (2011): 343–71.

13. Dyna Rochmyaningsih, "Massive Road Project Threatens New Guinea's Biodiversity," *Science* 374, no. 6565 (2021): 246–47.

14. Rhett Butler, "Tropical Forest's Lost Decade," *Mongabay*, December 17, 2019.

15. C. Garcia, "The Global Forest Transition as a Human Affair," *One Earth* 2 (2020): 417–28.
16. Holly Gibbs et al., "Tropical Forests Were the Primary Land Source for New Agricultural Lands in the 1980s and 1990s," *Proceedings of the National Academy of Science* 107, no. 38 (2010): 16732–37.
17. Karl Polanyi, *The Great Transformation* (New York: Farrar and Rinehart, 1944).
18. John McNeill, *The Great Acceleration: An Environmental History of the Anthropocene Since 1945* (Cambridge, MA: Harvard University Press, 2016).
19. Thomas Piketty and Emmanuel Saez, "Inequality in the Long Run," *Science* 344, no. 6186 (2014): 838–43.
20. Edward Wrigley, *Population and History* (London: Weidenfeld and Nicolson, 1969).
21. Zoe Laidlaw and Alan Lester, *Indigenous Communities and Settler Colonialism: Landholding, Loss, and Survival in an Interconnected World* (New York: Palgrave MacMillan, 2015).
22. Navin Ramankutty, Elizabeth Heller, and Jeanine Rhemtulla, "Prevailing Myths About Agricultural Abandonment and Forest Regrowth in the United States," *Annals of the Association of American Geographers* 100, no. 3 (2010): 502–12.
23. Patrick Meyfroidt, Thomas K. Rudel, and Eric Lambin, "Forest Transitions, Trade, and the Displacement of Land Use," *Proceedings of the National Academy of Sciences* 107, no. 49 (2010): 20917–22.
24. C. Hong et al., "Land Use Emissions Embodied in International Trade," *Science* 376, no. 6593 (2022): 597–603.
25. Ruth DeFries et al., "Planetary Opportunities: A Social Contract for Global Change Science to Contribute to a Sustainable Future," *Bioscience* 62, no. 6 (2012): 603–6.
26. Kemen Austin et al., "Trends in Size of Tropical Deforestation Events Signal Increasing Dominance of Industrial-Scale Drivers," *Environmental Research Letters* 5 (2017): 054009.
27. Saturnino Borras et al., "Towards a Better Understanding of Global Land Grabbing: An Editorial Introduction," *Journal of Peasant Studies* 38, no. 2 (2011): 209–16; Kyle Davis et al., "Accelerated Deforestation Driven by Large-Scale Land Acquisitions in Cambodia," *Nature Geosciences* 8, no. 10 (2015): 772–75.
28. Harvey Molotch, "The City as a Growth Machine: The Political Economy of Place," *American Journal of Sociology* 82, no. 2 (1976): 309–32; Thomas Rudel and Bruce Horowitz, *Tropical Deforestation: Small Farmers and Land Clearing in the Ecuadorian Amazon* (New York: Columbia University Press, 1993).
29. Gonzalo Aguirre Beltran, *Regions of Refuge* (Washington, DC: Society of Applied Anthropology, 1979).
30. R. M. Netting, *Smallholders, Householders: Farm Families and the Ecology of Intensive, Sustainable Agriculture* (Stanford, CA: Stanford University Press, 1993).
31. David B. Bray, *Mexico's Community Forest Enterprises: Success on the Commons and the Seeds of a Good Anthropocene* (Tucson: University of Arizona Press, 2020); Mark Poffenberger and Betsy McGean, *Village Voices, Forest Choices: Joint Forest Management in India* (Delhi: Oxford University Press 1996).

32. Here and in subsequent chapters, smallholders are defined as farming less than ten hectares of land, while the owners of large landholdings have more than twenty hectares of land.

33. Sarah Lowder, Jacob Skoet, and Terri Raney, "The Number, Size, and Distribution of Farms, Smallholder Farms, and Family Farms Worldwide," *World Development* 87 (2016): 16–29.

34. Leah Samberg et al., "Subnational Distribution of Average Farm Size and Smallholder Contributions to Global Food Production," *Environmental Research Letters* 11, no. 12 (2016): 124010.

35. T. S. Jayne, Jordan Chamberlin, and Derek Headey, "Land Pressures, the Evolution of Farming Systems, and Development Strategies in Africa: A Synthesis," *Food Policy* 48 (2014): 1–17; Alvaro D'Antona, Leah Van Wey, and Corey Hayashi, "Property Size and Land Cover Change in the Brazilian Amazon," *Population and Environment* 27 (2006): 373–96; Fernanda Michalski, Jean Metzger, and Carlos Peres, "Rural Property Size Drives Patterns of Upland and Riparian Forest Retention in a Tropical Deforestation Frontier," *Global Environmental Change* 20, no. 4 (2010): 705–12.

36. Gotz Schroth et al., "Conservation in Tropical Landscape Mosaics: The Case of the Cacao Landscape of Southern Bahia, Brazil," *Biodiversity and Conservation* 20 (2011): 1635–54.

37. Amy Lerner et al., "The Spontaneous Emergence of Silvo-pastoral Landscapes in the Ecuadorian Amazon: Patterns and Processes," *Regional Environmental Change* 15, no. 7 (2015): 1421–31.

38. Genevieve Michon et al., "Domestic Forests: A New Paradigm for Integrating Local Communities' Forestry into Tropical Forest Science," *Ecology and Society* 12, no. 2 (2007): 1.

39. Winkler, "Global Land Use Changes."

40. Yuta Masuda et al., "Emerging Research Needs and Policy Priorities for Advancing Land Tenure Security and Sustainable Development," in *Land Tenure Security and Sustainable Development*, ed. Margaret Holland, Yuta Masuda, and Brian Robinson (New York: Palgrave MacMillan, 2022), 313–26.

41. Patrick Meyfroidt et al., "Middle Range Theories of Land Use Change," *Global Environmental Change* 53 (2018): 52–67.

42. William Clark and Alicia Harley, "Sustainability Science: Towards a Synthesis," *Annual Review of Environment and Resources* 45 (2020): 331–86.

43. Polanyi, *The Great Transformation*.

44. Geoff Goodwin, "Rethinking the Double Movement: Expanding the Frontiers of Polanyian Analysis in the Global South," *Development and Change* 49, no. 5 (2018): 1268–90.

45. Steven Brechin and William Fenner, "Karl Polanyi's Environmental Sociology: A Primer," *Environmental Sociology* 3, no. 3 (2017): 1–10.

46. John Meyer et al., "World Society and the Nation State," *American Journal of Sociology* 103, no. 1 (1997): 144–81.

47. Sven Wunder et al., "REDD+ in Theory and Practice: How Local Projects Can Inform Jurisdictional Approaches," *Frontiers in Forests and Global Change* 3, no. 11 (2020), https://www.cifor.org/publications/pdf_files/articles/AWunder2001.pdf.

48. Daniel Muller et al., "Regime Shifts Limit the Predictability of Land-System Change," *Global Environmental Change* 28 (2014): 75–83.
49. Émile Durkheim quoted in Wolfgang Streeck and Lane Kenworthy, "Theories and Practices of Neo-corporatism," in *The Handbook of Political Sociology: States, Civil Societies, and Globalization*, ed. T. Janoski et al. (Cambridge: Cambridge University Press, 2005), 441–60, at 445.
50. Lucy Tompkins, "What Does It Mean to Sell Forest Candy?," *New York Times*, December 5, 2021.
51. Philippe Schmitter, "Still a Century of Corporatism?," *Review of Politics* 36, no. 1 (1974): 85–131; Oscar Molina and Martin Rhodes, "Corporatism: The Past, Present, and Future of a Concept," *Annual Review of Political Science* 5 (2002): 305–31; John Dryzek et al., "The Environmental Transformation of the State: The USA, Norway, Germany, and the UK," *Political Studies* 50 (2002): 659–82; Thomas K. Rudel, "Shocks, States, and Societal Corporatism: A Shorter Path to Sustainability?," *Journal of Environmental Studies and Sciences* 9, no. 4 (2019): 429–36.
52. Perola Oberg et al., "Disrupted Exchange and Declining Corporatism," *Government and Opposition* 46, no. 3 (2011): 365–91.
53. Schmitter, "Still a Century."
54. Robert Dahl, *Who Governs? Democracy and Power in an American City* (New Haven, CT: Yale University Press, 1961).
55. Molina and Rhodes, "Corporatism."
56. Beth McKillen, "The Corporatist Model, World War I, and the Debate About the League of Nations," *Diplomatic History* 15, no. 2 (1991): 171–97.
57. Schmitter, "Still a Century."
58. Frances Seymour and Jonas Busch, *Why Forests? Why Now? The Science, Economics, and Politics of Tropical Forests and Climate Change* (Washington, DC: Center for Global Development, 2016).
59. John W. Reid and Thomas Lovejoy, *Ever Green: Saving Big Forests to Save the Planet* (New York: Norton, 2022).
60. Chris Barrett et al., "Conserving Tropical Biodiversity amid Weak Institutions," *Bioscience* 51, no. 6 (2001): 497–502.
61. Schmitter, "Still a Century."
62. Thomas K. Rudel, *Defensive Environmentalists and the Dynamics of Global Reform* (New York: Cambridge University Press, 2013).
63. Arild Angelsen, "REDD+ as Result-based Aid: General Lessons and Bilateral Agreements of Norway," *Review of Development Economics* 21, no. 2 (2017): 237–64.
64. Wunder, "REDD+."
65. W. Sunderlin et al., "Creating an Appropriate Tenure Foundation for REDD+: The Record to Date and Prospects for the Future," *World Development* 106 (2018): 176.
66. Arild Angelsen et al., "Learning from REDD+: A Response to Fletcher et al.," *Conservation Biology* 31, no. 3 (2017): 718–20.
67. Sunderlin et al., "REDD+."
68. Sunderlin et al., "REDD+."
69. Tomas Sikor and Jacopo Baggio, "Can Smallholders Engage in Tree Plantations? An Entitlements Analysis from Vietnam," *World Development* 64 (2014): 101–12.
70. Hans Nicholas Jong, "RSPO Fails to Deliver on Environmental and Social Sustainability, Study Finds," *Mongabay*, July 11, 2018.

71. Diane Bates and Thomas K. Rudel, "The Political Ecology of Tropical Rain Forest Conservation: A Cross-National Analysis," *Society and Natural Resources* 13, no. 7 (2000): 587–603.
72. Thomas K. Rudel, "Indigenous-Driven Sustainability Initiatives in Mountainous Regions: The Shuar in the Tropical Andes of Ecuador," *Mountain Research and Development* 41, no. 1 (2021): 22–28.
73. Hannah Ritchie and Max Roser, "Protected Areas and Conservation," Our World in Data, 2021, https://ourworldindata.org/biodiversity.
74. FAO, *Global Forest Resources Assessment, 2020.*
75. Doug Boucher, Sarah Roquemore, and Estrellita Fitzhugh, "Brazil's Success in Reducing Deforestation," *Tropical Conservation Science* 6, no. 3 (2013): 426–45.
76. Christopher Wolf et al., "A Forest Loss Report Card for the World's Protected Areas," *Nature Ecology and Evolution* 5, no. 4 (2021): 520–29.
77. Greg Asner et al., "A Contemporary Assessment of Change in Humid Tropical Forests," *Conservation Biology* 23 (2009): 1386–95.
78. François Ruf, "Tree Crops as Deforestation and Reforestation Agents: The Case of Cocoa in Côte d'Ivoire and Sulawesi," in *Agricultural Technologies and Tropical Deforestation*, ed. A. Angelsen and D. Kaimowitz (Wallingford, UK: CABI Publishing, 2001), 291–315.
79. Ester Boserup, *The Conditions of Agricultural Growth: The Economics of Agrarian Change Under Population Pressure* (London: Allen and Unwin, 1965).
80. Virginia Garcia et al., "Agricultural Intensification and Land Use Change: Assessing Country-Level Induced Intensification, Land Sparing and Rebound Effect," *Environmental Research Letters* 15, no. 8 (2020): 085007.
81. Alexander Mather and Carolyn Needle, "The Forest Transition: A Theoretical Basis," *Area* 30, no. 2 (1998): 117–24.
82. Sarah Philips, *This Land, This Nation: Conservation, Rural America, and the New Deal* (New York: Cambridge University Press, 2007).
83. John Schelhas, Tom Brandeis, and Thomas K. Rudel, "Reforestation and Natural Regeneration in Forest Transitions: Patterns and Implications from the U.S. South," *Regional Environmental Change* 21, no. 1 (2021): 8; David Kaczan, "Can Roads Contribute to Forest Transitions?," *World Development* 129 (2020): 104898.
84. Ruf, "Tree Crops."
85. Andrew George and Alexander Bennett, *Case Studies and Theory Development in the Social Sciences* (Cambridge, MA: MIT Press, 2005).

3. Forest Losses, the Conservation Movement, and Protected Areas

1. Edward O. Wilson, *Half-Earth: Our Planet's Fight for Life* (New York: Norton, 2016).
2. Food and Agricultural Organization (FAO) and United Nations Environmental Program (UNEP), *The State of the World's Forests 2020: Forests, Biodiversity and People* (Rome: Food and Agricultural Organization of the United Nations, 2020).

3. Thomas K. Rudel, "Indigenous-Driven Sustainability Initiatives in Mountainous Regions: The Shuar in the Tropical Andes of Ecuador," *Mountain Research and Development* 41, no. 1 (2021): 22–28.
4. Dyna Rochmyaningsih, "Massive Road Project Threatens New Guinea's Biodiversity," *Science* 374, no. 6565 (2021): 246–47.
5. C. Vancutsem et al., "Long-Term (1990–2019) Monitoring of Forest Cover Changes in the Humid Tropics," *Science Advances* 7 (2021): eabe1603.
6. Philip Curtis et al., "Classifying Drivers of Global Forest Loss," *Science* 361 (2018): 1108–11.
7. FAO, *Global Forest Resources Assessment, 2020* (Rome: Food and Agriculture Organization of the United Nations, 2020).
8. Emmanuel Chidumayo and Davison Gumbo, *The Dry Forests of Africa: Managing for Products and Services* (London: Earthscan, 2010).
9. Elsa Ordway, Greg Asner, and Eric Lambin, "Deforestation Risk Due to Commodity Crop Expansion in Sub-Saharan Africa," *Environmental Research Letters* 12 (2017): 044015.
10. FAO, *Global Forest Resources Assessment, 2020.*
11. John Terborgh, *Requiem for Nature* (Washington, DC: Island Press, 1999).
12. Ronald Foresta, *Amazon Conservation During the Age of Development* (Gainesville: University of Florida Press, 1991).
13. Frances Seymour and Jonas Busch, *Why Forests? Why Now? The Science, Economics, and Politics of Tropical Forests and Climate Change* (Washington, DC: Center for Global Development, 2016).
14. FAO, *Global Forest Resources Assessment, 2020.*
15. Rhett Butler, "Tropical Forest's Lost Decade," *Mongabay*, December 17, 2019.
16. Patricia Widener, *Oil Injustice: Resisting and Conceding a Pipeline in Ecuador* (Lanham, MD: Rowman & Littlefield, 2011).
17. David Foster, *Wildlands and Woodlands: Farmlands and Communities—Broadening the Vision for New England* (Petersham, MA: Harvard Forest, 2017).
18. Thomas Rudel et al., "From Middle to Upper Class Sprawl? Land Use Controls and Real Estate Development in Northern New Jersey," *Annals of the Association of American Geographers* 101 (2011): 609–24.
19. Thomas K. Rudel, *Defensive Environmentalists and the Dynamics of Global Reform* (New York: Cambridge University Press, 2013).
20. Christopher Wolf et al., "A Forest Loss Report Card for the World's Protected Areas," *Nature Ecology and Evolution* 5, no. 4 (2021): 520–29.
21. John Schelhas and Max Pfeffer, *Saving Forests, Protecting People: Environmental Conservation in Central America* (Lanham, MD: Altamira Press, 2008).
22. Deborah Barry and Ruth Meinzen-Dick, "The Invisible Map: Community Tenure Rights," in *The Social Lives of Forests: The Past, Present, and Future of Woodland Resurgence*, ed. Kathleen Morrison, Christine Padoch, and Susanna Hecht (Chicago: University of Chicago Press, 2014), 291–302.
23. S. Garnett et al., "A Spatial Overview of the Global Importance of Indigenous Lands for Conservation," *Nature Sustainability* 1 (2018): 369–74.
24. J. Fa et al., "Importance of Indigenous Peoples' Lands for the Conservation of Intact Forest Landscapes," *Frontiers in Ecology and the Environment* 18, no. 3

(2020): 135–40; FAO-UNEP, *Forests, Biodiversity, and People*; Barry and Meinzen-Dick, "The Invisible Map."

25. Eric Wakker, *The Kalimantan Border Oil Palm Mega-Project* (Amsterdam: Friends of the Earth—Netherlands, 2006.)

26. Tomas Sikor et al., *Upland Transformations in Vietnam* (Singapore: National University of Singapore Press, 2011).

27. Manuela Andreoni, Hiroko Tabuchi, and Albert Sun, "Destroying the Amazon for Leather Auto Seats," *New York Times*, November 18, 2021.

28. Butler, "Lost Decade."

29. Igarape Institute, *Connecting the Dots: Territories and Trajectories of Environmental Crime in the Brazilian Amazon and Beyond* (Rio de Janeiro: Igarape Institute, 2022).

30. Anne M. Larson and Ganga Dahal, "Forest Tenure Reform: New Resource Rights for Forest-Based Communities," *Conservation and Society* 10, no. 2 (2012): 77–90.

31. David B. Bray, *Mexico's Community Forest Enterprises: Success on the Commons and the Seeds of a Good Anthropocene* (Tucson: University of Arizona Press, 2020).

32. Bray, *Community Forest Enterprises*.

33. Bray, *Community Forest Enterprises*.

34. Bray, *Community Forest Enterprises*.

35. Jennifer Alix-Garcia et al., "Payments for Environmental Services Supported Social Capital While Increasing Land Management," *PNAS* 115 (2018): 7016–21.

36. Arild Angelsen, "REDD+ as Result-Based Aid: General Lessons and Bilateral Agreements of Norway," *Review of Development Economics* 21, no. 2 (2017): 237–64.

37. Jens Lund and Thorsten Treue, "Are We Getting There? Evidence of Decentralized Forest Management from the Tanzanian Miombo Woodlands," *World Development* 36, no. 12 (2008): 2780–800; Jens Lund, Rebecca Rett, and Jesse Ribot, "Trends in Research on Forestry Decentralization Policies," *Current Opinion in Environmental Sustainability* 32, no. 1 (2018): 17–22.

38. Angelsen, "REDD+ as Result-Based Aid."

39. Hans Nicholas Jong, "Indonesia Terminates Agreement with Norway on $1b REDD+ scheme," *Mongabay*, September 10, 2021.

40. Lee Alston and Bernardo Mueller, "Legal Reserve Requirements in Brazilian Forests: Path Dependent Evolution of De Facto Legislation," *Revista Economia* 8, no. 4 (2007): 25–53.

41. Alston and Mueller, "Legal Reserve Requirements."

42. Fernanda Michalski, Jean Metzger, and Carlos Peres, "Rural Property Size Drives Patterns of Upland and Riparian Forest Retention in a Tropical Deforestation Frontier," *Global Environmental Change* 20, no. 4 (2010): 705–12.

43. Doug Boucher, Sarah Roquemore, and Estrellita Fitzhugh, "Brazil's Success in Reducing Deforestation," *Tropical Conservation Science* 6, no. 3 (2013): 426–45.

44. Yann Polain de Waroux et al., "Rents, Actors, and the Expansion of Commodity Frontiers in the Gran Chaco," *Annals of the American Association of Geographers* 108, no. 1 (2018): 204–25.

45. Yann Polain de Waroux et al., "Land-Use Policies and Corporate Investments in Agriculture in the Gran Chaco and Chiquitano," *Proceedings of the National Academy of Sciences* 113, no. 15 (2016): 4021–26.

46. Yann Polain de Waroux et al., "The Restructuring of South American Soy and Beef Production and Trade Under Changing Environmental Regulations," *World Development* 121 (2019): 188–202.
47. Polain de Waroux et al., "Restructuring of Soy and Beef Production."
48. Hans Nicholas Jong, "Indonesia Forest Clearing Ban Is Made Permanent, but Labelled Propaganda," *Mongabay*, August 14, 2019.
49. Jong, "Forest Clearing Ban."
50. Fred Pearce, *A Trillion Trees: How We Can Reforest Our World* (London: Granta, 2021).
51. L. Gatti et al., "Amazonia as a Carbon Source linked to Deforestation and Climate Change," *Nature* 595, no. 7867 (2021): 388–93.
52. Gatti, "Amazonia as a Carbon Source."
53. Seymour and Busch, *Why Forests, Why Now?*

4. Rural-Urban Migration, Land Abandonment, and the Spread of Secondary Forests

1. Ruth Chazdon, *Second Growth: The Promise of Tropical Forest Regeneration in an Age of Deforestation* (Chicago: University of Chicago Press, 2014).
2. B. Bernal, L. Murray, and T. Pearson, "Carbon Dioxide Removal Rates from Forest Landscape Restoration Activities," *Carbon Balance Management* 13 (2018): 22.
3. Simon L. Lewis et al., "Regenerate Natural Forests to Store Carbon," *Nature* 568 (2019): 25–28.
4. David Foster and John Aber, *Forests in Time: The Environmental Consequences of 1000 Years of Change in New England* (New Haven, CT: Yale University Press, 2004).
5. James Goldthwait, "A Town That Has Gone Downhill," *Geographical Review* 17, no 4. (1927): 527–52; Hugh Raup, "The View from John Sanderson's Farm: A Perspective for the Use of Land," *Forest History* 10, no. 1 (1966): 2–11.
6. Alexander Mather and Carolyn Needle, "The Forest Transition: A Theoretical Basis," *Area* 30, no. 2 (1998): 117–24.
7. Goldthwait, "Town That Has Gone Downhill."
8. Raup, "The View from John Sanderson's Farm."
9. Foster and Aber, *Forests in Time*.
10. Goldthwait, "Town That Has Gone Downhill."
11. Mark Drummond and Thomas Loveland, "Land Use Pressure and a Transition to Forest-Cover Loss in the Eastern United States," *BioScience* 60, no. 4 (2010): 286–98.
12. Tiago Oliviera et al., "Is Portugal's Forest Transition Going Up in Smoke?," *Land Use Policy* 66 (2017): 214–26.
13. Jeffrey Bentley, *Today There Is No Misery: The Ethnography of Farming in Northwest Portugal* (Tucson: University of Arizona Press, 1992.)
14. Bentley, *Today There Is No Misery.*
15. Bentley, *Today There Is No Misery.*

16. Jeffrey Bentley, "Bread Forests and New Fields: The Ecology of Reforestation and Forest Clearing Among Small-Woodland Owners in Portugal," *Journal of Forest History* 33, no. 4 (1989): 188–95.

17. Mather and Needle, "Forest Transition."

18. Bentley, *Today There Is No Misery.*

19. B. J. Woodruffe, "Rural Land Use Planning in West Germany," in *Rural Land Use Planning in Developed Nations*, ed. Paul J. Cloke (London: Unwin Hyman 1989), 104–29.

20. Patrick Meyfroidt, Thomas K. Rudel, and Eric Lambin, "Forest Transitions, Trade, and the Displacement of Land Use," *Proceedings of the National Academy of Sciences* 107, no. 49 (2010): 20917–22.

21. Oliviera et al., "Portugal's Forest Transition."

22. Arthur Raper and Ira Reid, *Sharecroppers All* (New York: Russell & Russell, 1941).

23. Howard Odum, *The Southern Regions of the United States* (Chapel Hill: The University of North Carolina Press, 1936).

24. Arthur Raper, *A Preface to Peasantry: A Tale of Two Black Belt Counties* (Chapel Hill: University of North Carolina Press, 1936); Philip Beck and Milton Forster, *Six Rural Problem Areas: Relief, Resources, and Rehabilitation* (Washington, DC: Works Progress Administration, 1935).

25. Isabel Wilkerson, *The Warmth of Other Suns: The Epic Story of America's Great Migration* (New York: Random House, 2010).

26. Raper, *A Preface to Peasantry.*

27. Robert S. McElvaine, *The Great Depression: America, 1929–1941* (New York: Times Books, 1984).

28. Gilbert C. Fite, *Cotton Fields No More: Southern Agriculture, 1865–1980* (Lexington: University of Kentucky Press, 1983).

29. Theodore Saloutos, *The American Farmer and the New Deal* (Ames: Iowa State University Press, 1982).

30. Thomas K. Rudel, "Did a Green Revolution Reforest the American South?," in *Agricultural Technologies and Tropical Deforestation*, ed. A. Angelsen and D. Kaimowitz (New York: CABI Press, 2001), 33–54.

31. Karen O'Neill, *Rivers by Design: State Power and the Origins of U.S. Flood Control* (Durham, NC: Duke University Press, 2006).

32. Sonja Oswalt et al., *Forest Resources of the United States* (Washington, DC: United States Department of Agriculture, 2017); Mark Drummond and Thomas Loveland. "Land Use Pressure and a Transition to Forest-Cover Loss in the Eastern United States," *BioScience* 60, no. 4 (2010): 286–98.

33. Amy Daniels, "Forest Expansion in Northwest Costa Rica: Conjuncture of the Global Market, Land Use Intensification, and Forest Protection," in *Reforesting Landscapes: Linking Pattern and Process*, ed. H. Nagendra and J. Southworth (Dordrecht: Springer, 2010), 227–52.

34. Daniels, "Forest Expansion."

35. Marc Edelman, "Extensive Land Use and the Logic of the Latifundio: A Case Study in Guanacaste Province, Costa Rica," *Human Ecology* 13, no. 2 (1985): 153–85.

36. Daniels, "Forest Expansion."

37. Daniels, "Forest Expansion."

38. Daniels, "Forest Expansion."
39. Daniels, "Forest Expansion."
40. Daniels, "Forest Expansion."
41. Karen Allen and Steve Padgett Vasquez, "Forest Cover, Development, and Sustainability in Costa Rica: Can One Policy Fit All?," *Land Use Policy* 67 (2017): 212–21.
42. Isaline Jadin, Patrick Meyfroidt, and Eric Lambin, "International Trade and Land Use Intensification and Spatial Reorganization Explain Costa Rica's Forest Transition," *Environmental Research Letters* 11 (2016): 035005.
43. J. Leighton Reid et al., "The Ephemerality of Secondary Forests in Southern Costa Rica," *Conservation Letters* 12 (2019): e12607.
44. Mather and Needle, "Forest Transition."
45. Jadin, Meyfroidt, and Lambin, "International Trade and Land Use."
46. Johan Oldekop et al., "International Migration Drives Reforestation in Nepal," *Global Environmental Change* 52 (2018): 66–74; Susanna Hecht et al., "People in Motion, Forests in Transition: Trends in Migration, Urbanization, and Remittances and Their Effects on Tropical Forests" (Bogor, Indonesia: CIFOR, 2015); Bradley Walters, "Migration, Land Use, and Forest Change in St. Lucia, West Indies," *Land Use Policy* 51 (2016): 290–300; Bradley Walters, *The Greening of Saint Lucia: Economic Development and Environmental Change in the Eastern Caribbean* (Kingston, Jamaica: University of the West Indies Press, 2019).
47. Walters, "Migration, Land Use, and Forest Change"; Walters, *The Greening of Saint Lucia.*
48. Johan Von Thünen, *Von Thünen's Isolated State*, trans. Carla M. Wartenberg (London: Pergamon Press, 1967).
49. S. Collins et al., "An Integrated Conceptual Framework for Long-Term Social–Ecological Research," *Frontiers in Ecology and the Environment* 9 (2011): 351–57.
50. Virginia Garcia et al., "Agricultural Intensification and Land Use Change: Assessing Country-Level Induced Intensification, Land Sparing, and Rebound Effect," *Environmental Research Letters* 15, no. 8 (2020): 085007.
51. Thomas Hertel, Navin Ramankutty, and Uris Baldos, "Global Market Integration Increases Likelihood That a Future African Green Revolution Could Increase Cropland Use and CO_2 Emissions," *Proceedings of the National Academy of Sciences* 111, no. 38 (2014): 13799–804.
52. Meyfroidt, Rudel, and Lambin, "Forest Transitions."
53. Thomas Rudel and Bruce Horowitz, *Tropical Deforestation: Small Farmers and Land Clearing in the Ecuadorian Amazon* (New York: Columbia University Press, 1993).
54. T. Tseng et al., "Influence of Land Tenure Interventions on Human Well-Being and Environmental Outcomes," *Nature Sustainability* 4 (2021): 242–51.
55. Fernanda Michalski, Jean Metzger, and Carlos Peres, "Rural Property Size Drives Patterns of Upland and Riparian Forest Retention in a Tropical Deforestation Frontier," *Global Environmental Change* 20, no. 4 (2010): 705–12; Alvaro D'Antona, Leah Van Wey, and Corey Hayashi, "Property Size and Land Cover Change in the Brazilian Amazon," *Population and Environment* 27 (2006): 373–96.
56. T. S. Jayne, Jordan Chamberlin, and Derek Headey, "Land Pressures, the Evolution of Farming Systems, and Development Strategies in Africa: A Synthesis," *Food Policy* 48 (2014): 1–17.

57. Norman Borlaug, "Feeding a Hungry World," *Science* 318 (2007): 359.

58. Ben Phalan et al., "Reconciling Food Production and Biodiversity Conservation: Land Sharing and Land Sparing Compared," *Science* 333 (2011): 1289–91; Fred Pearce, "Sparing Versus Sharing: The Great Debate Over How to Protect Nature," *Environment 360*, 2018, https://e360.yale.edu/features/sparing-vs-sharing-the-great-debate-over-how-to-protect-nature.

59. Daniel Redo et al., "Asymmetric Forest Transition Related to the Interaction of Socio-economic Development and Forest Type in Central America," *Proceedings of the National Academy of Sciences* 109, no. 23 (2012): 8839–44; Ernesto Feder, *The Rape of the Peasantry: Latin America's Landholding System* (New York: Anchor Books, 1971).

60. Leslie Duram, Jon Bathgate, and Christina Ray, "A Local Example of Land Use Change: Southern Illinois, 1807, 1938, 1993," *Professional Geographer* 56, no. 1 (2004): 127–40.

61. Nyein Chan and Shinya Takeda, "The Transition Away from Swidden Agriculture and Trends in Biomass Accumulation in Fallow Forests: Case Studies in the Southern Chin Hills of Myanmar," *Mountain Research and Development* 36, no. 3 (2016): 320–31; Redo et al., "Asymmetric Forest Transition"; Tobias Kuemmerle et al., "Hotspots of Land Use Change in Europe," *Environmental Research Letters* 11 (2016): 064020; A. Sofia Nanni and H. Ricardo Grau, "Agricultural Adjustment, Population Dynamics and Forest Redistribution in a Subtropical Watershed of NW Argentina," *Regional Environmental Change* 14 (2014): 1641–49.

62. Alywn Gentry, "Patterns of Diversity and Floristic Composition in Neotropical Montane Forests," in *Biodiversity and Conservation of Neo-tropical Montane Forests*, ed. S. Churchill (New York: New York Botanical Garden, 1995), 103–26.

63. Thomas Hertel, "Growing Food and Protecting Nature Don't Have to Conflict— Here's How They Can Work Together," *The Conversation*, March 9, 2021, https://theconversation.com/growing-food-and-protecting-nature-dont-have-to-conflict-heres-how-they-can-work-together-146069.

64. Alexander Mather, J. Fairbairn, and C. Needle, "The Course and Drivers of the Forest Transition: The Case of France," *Journal of Rural Studies* 15, no. 1(1999): 65–90; Alexander Mather and J. Fairbairn, "From Floods to Reforestation: The Forest Transition in Switzerland," *Environment and History* 6, no. 4 (2000): 399–421.

65. Navin Ramankutty and Oliver Coomes, "Land-use Regime Shifts: An Analytical Framework and Agenda for Future Land Use Research," *Ecology and Society* 21, no. 2 (2016): 1.

66. Kuemmerle et al., "Hotspots."

67. Stephan Estel et al., "Mapping Farmland Abandonment and Recultivation Across Europe Using MODIS NDVI Time Series," *Remote Sensing of the Environment* 163 (2015): 312–25.

68. Angel Leyva Galan, "Metodología para el desarollo de la biodiversidad vegetal," in *Agroecologia en el tropico: Ejemplos de Cuba*, ed. A. Leyva and J. Pohlin (Aachen: Shaker Verlag, 2005), 165–82.

69. Thomas K. Rudel, *Shocks, States, and Sustainability: The Origins of Radical Environmental Reforms* (New York: Oxford University Press, 2019).

70. Naomi Schwartz et al., "Reversals of Reforestation Across Latin America Limit Climate Mitigation Potential of Tropical Forests," *Frontiers in Forests and Global Change* 3 (2020): 85; J. Leighton Reid et al., "The Ephemerality of Secondary Forests in Southern Costa Rica," *Conservation Letters* 12 (2019): e12607.
71. F. Elias et al., "Assessing the Growth and Climate Sensitivity of Secondary Forests in Climate Sensitive Amazonian Landscapes," *Ecology* 101, no. 3 (2020): ecy2954.
72. Lewis et al., "Secondary Forests."

5. Planted Forests

1. A. Ziegler et al., "Carbon Outcomes of Major Land-Cover Transitions in SE Asia: Great Uncertainties and REDD+ Policy Implications," *Global Change Biology* 18 (2012): 3087–99; B. Bernal, L. Murray, and T. Pearson, "Carbon Dioxide Removal Rates from Forest Landscape Restoration Activities," *Carbon Balance Management* 13 (2018): 22.
2. Alexander Mather, "Forest Transition Theory and the Reforestation of Scotland," *Scottish Geographical Journal* 120 (2004): 83–98.
3. Peter Dauvergne, *Shadows in the Forest: Japan and the Politics of Timber in Southeast Asia* (Cambridge, MA: MIT Press, 1997).
4. Greg Asner et al., "A Contemporary Assessment of Change in Humid Tropical Forests," *Conservation Biology* 23 (2009): 1386–95.
5. Mather, "Forest Transition Theory."
6. Sean Sloan et al., "The Forest Transformation: Planted Tree Cover and Regional Dynamics of Tree Gains and Losses," *Global Environmental Change* 59 (2019): 101988.
7. FAO, *Global Forest Resources Assessment, 2020* (Rome: Food and Agriculture Organization of the United Nations, 2020).
8. Paulo Charruadas and Chloe Deligne, "Cities Hiding the Forests: Wood Supply, Hinterlands, and Urban Agency in the Southern Low Countries, Thirteen—Eighteenth Centuries," in *Urbanizing Nature: Actors and Agency (Dis)Connecting Cities and Nature Since 1500*, ed. T. Soens et al. (London: Taylor and Francis, 2019), 112–34.
9. Eduardo Brondizio, *The Amazonian Caboclo and the Açaí Palm: Forest Farmers in the Global Market* (New York: New York Botanical Garden, 2008).
10. John Schelhas, Tom Brandeis, and Thomas K. Rudel, "Reforestation and Natural Regeneration in Forest Transitions: Patterns and Implications from the U.S. South," *Regional Environmental Change* 21, no. 1 (2021): 260–69.
11. David Kaczan, "Can Roads Contribute to Forest Transitions?," *World Development* 129 (2020): 104898.
12. M. E. Fagan et al., "The Expansion of Tree Plantations Across Tropical Biomes," *Nature Sustainability* 5 (2021): 681–88.
13. Sandra Baptista and Thomas K. Rudel, "Is the Atlantic Forest Re-emerging? Urbanization, Industrialization, and the Forest Transition in Santa Catarina, Southern Brazil," *Environmental Conservation* 33, no. 3 (2006): 195–202.

14. Sloan et al., "Forest Transformation."
15. Gould Kenneth, David Pellow, and Allan Schnaiberg, *Treadmill of Production: Injustice and Unsustainability in the Global Economy* (Boulder, CO: Paradigm Publishers, 2008).
16. Andrew P. Vayda and Ahmad Sahur, "Forest Clearing and Pepper Farming by Bugis Migrants in East Kalimantan: Antecedents and Impacts," *Indonesia* 39, no. 4 (1985): 93–110.
17. John W. Reid and Thomas Lovejoy, *Ever Green: Saving Big Forests to Save the Planet* (New York: Norton, 2022).
18. P. Potapov et al., "The Last Frontiers of Wilderness: Tracking Loss of Intact Forest Landscapes from 2000 to 2013," *Science Advances* 3, no. 1 (2017): 1600821.
19. Reid and Lovejoy, *Ever Green*.
20. Jai Chen, John Zinda, and E. Ting Yeh, "Recasting the Rural: State, Society, and Environment in Contemporary China," *Geoforum* 78 (2017): 83–88.
21. John Zinda et al., "Dual Function Forests in the Returning Forests to Farmland Program and Flexibility in Environmental Policy in China," *Geoforum* 78 (2017): 119–32.
22. Michael Wolosin, *Large-Scale Forestation for Climate Mitigation: Lessons from South Korea, China, and India* (Washington, DC: Climate and Land Use Alliance, 2017).
23. James Scott, *Seeing Like a State: How Certain Schemes to Improve the Human Condition Have Failed* (New Haven, CT: Yale University Press, 1998); Chen, Zinda, and Yeh, "Recasting the Rural State."
24. Zinda et al., "Dual Function Forests."
25. Jing Chen et al., "Vegetation Structural Change Since 1981 Significantly Enhanced the Terrestrial Carbon Sink," *Nature Communications* 10 (2019): 4259.
26. Pam McElwee, *Forests Are Gold: Trees, People, and Environmental Rule in Vietnam* (Seattle: University of Washington Press, 2016).
27. Patrick Meyfroidt and Eric Lambin, "The Causes of the Reforestation in Vietnam," *Land Use Policy* 25, no. 2 (2008): 182–97.
28. Tomas Sikor and Jacopo Baggio, "Can Smallholders Engage in Tree Plantations? An Entitlements Analysis from Vietnam," *World Development* 64 (2014): 101–12.
29. Sikor and Baggio, "Smallholder Tree Plantations."
30. McElwee, *Forests Are Gold*.
31. Dao Truong, Masayuki Yanagisawa, and Yasuyuki Kono, "Forest Transition in Vietnam: A Case Study of Northern Mountain Region," *Forest Policy and Economics* 76 (2017): 72–80.
32. Sikor and Baggio, "Smallholder Tree Plantations"; Meyfroidt and Lambin, "Causes of Reforestation."
33. David Preston, "Changed Household Livelihood Strategies in the Cordillera of Luzon," *Tijdschrift voor Economische en Sociale Geografie* 89, no. 4 (1998): 371–83.
34. Meyfroidt and Lambin, "Causes of Reforestation."
35. Sikor and Baggio, "Smallholder Tree Plantations."
36. FAO, *Global Forest Resources Assessment, 2010* (Rome: Food and Agriculture Organization of the United Nations, 2010).
37. Alexander Mather, "Recent Asian Transitions in Relation to Forest Transition Theory," *International Forestry Review* 9, no. 1 (2007): 491–502.

38. Guillame Lestrelin, Jean Cristophe Castella, and Jeremy Bourgoin, "Territorial-izing Sustainable Development: The Politics of Land-Use Planning in Laos," *Journal of Contemporary Asia* (2012): 1–22.

39. Thilde Bruun et al., "Environmental Consequences of the Demise in Swidden Cultivation in Southeast Asia: Carbon Storage and Soil Quality," *Human Ecology* 37 (2009): 375–88.

40. Lestrelin, "Territorializing Sustainable Development"; Keith Barney, "China and the Production of Forest-Lands in Lao PDR: A Political Ecology of Transnational Enclosure," in *Taking Southeast Asia to Market: Commodities, Nature, and People in the Neoliberal Age*, ed. Joseph Nevins and Nancy Peluso (Ithaca, NY: Cornell University Press, 2008), 91–107.

41. Alan Ziegler, Jefferson Fox, and Jiancho Xu, "The Rubber Juggernaut," *Science* 324 (2009): 1024–25.

42. Christopher Barr and Jeffrey Sayer, "The Political Economy of Reforestation and Forest Restoration in Asia-Pacific: Critical Issues for REDD+," *Biological Conservation* 154 (2012): 9–19; Dauvergne, *Shadows in the Forest*.

43. Rosamond Naylor et al., "Decentralization and the Environment: Assessing Smallholder Oil Palm Development in Indonesia," *Ambio* 48 (2019): 1195–1208; Frances Seymour and Jonas Busch, *Why Forests? Why Now? The Science, Economics, and Politics of Tropical Forests and Climate Change* (Washington, DC: Center for Global Development, 2016).

44. Alain Rival and Patrice Levang, *Palms of Controversies: Oil Palm and Development Challenges* (Bogor, Indonesia: CIFOR, 2014).

45. Directorate General of Estate Crops, "Tree Crop Estate Statistics of Indonesia, Palm Oil, 2015–2017" (Jakarta, Indonesia: Department of Agriculture, 2016).

46. Rival and Levang, *Palms of Controversies.*

47. Naylor et al., "Decentralization."

48. Rival and Levang, *Palms of Controversies.*

49. Florencia Pulhin, Rodel Lasco, and Joan Urquiola, "The Carbon Sequestration Potential of Oil Palm in Bohol, Philippines," *Ecosystems and Development Journal* 4, no. 2 (2014): 14–19.

50. Sloan, "Forest Transformation"

51. Bruun et al., "Demise of Swidden Cultivation."

52. Daniela Miteva, Colby Loucks, and Subhrendu Pattanayak, "Social and Environmental Impacts of Forest Management Certification in Indonesia," *Plos One* 10, no. 7 (2015): e0129675.

53. Kimberly Carlson et al., "Effect of Oil Palm Sustainability Certification on Deforestation and Fire in Indonesia," *Proceedings of the National Academy of Sciences* 115, no. 1 (2018): 121–26.

54. Schelhas, Brandeis, and Rudel, "Reforestation and Natural Regeneration."

55. Schelhas, Brandeis, and Rudel, "Reforestation and Natural Regeneration."

56. Huan Gu et al., "The Carbon Balance of the Southeastern U.S. Forest Sector as Driven by Recent Disturbance Trends," *Journal of Geophysical Research: Biogeosciences* 124 (2019): 2786–803.

57. Schelhas, Brandeis, and Rudel, "Reforestation and Natural Regeneration."

58. Sloan, "Forest Transformation."

59. Gu et al., "Carbon Balance."

60. Fred Pearce, *A Trillion Trees: How We Can Reforest Our World* (London: Granta, 2021).
61. The sources and estimates in table 5.1, with the exception of the FAO source for the 2020 estimate, come from Robert Heilmayr et al., "A Plantation-Dominated Forest Transition in Chile," *Applied Geography* 75 (2016): 71–82.
62. Robert Heilmayr, Cristian Echeverria, and Eric Lambin, "Impacts of Chilean Forest Subsidies on Forest Cover, Carbon and Biodiversity," *Nature—Sustainability* 3 (2020): 701–9.
63. Heilmayr et al., "A Plantation-Dominated Forest Transition."
64. Heilmayr, Echeverria, and Lambin, "Impacts of Chilean Forest Subsidies"; Heilmayr et al., "A Plantation-Dominated Forest Transition."
65. Robert Heilmayr and Eric Lambin, "Impacts of Nonstate, Market-Driven Governance on Chilean Forests," *Proceedings of the Natural Academy of Sciences* 113, no. 11 (2016): 2910–15.
66. Heilmayr and Lambin, "Impacts."
67. Heilmayr and Lambin, "Impacts."
68. Heilmayr and Lambin, "Impacts."
69. America Duran and Olga Barbosa, "Seeing Chile's Forest for the Tree Plantations," *Science* 365, no. 6460 (2019): 1388.
70. United Nations, Climate Secretariat, *Intended Nationally Determined Contributions* (New York: United Nations, 2016).
71. Lewis et al., "Secondary Forests."
72. Joel Migdal, *Strong Societies and Weak States: State-Society Relationships and State Capabilities in the Developing World* (Princeton, NJ: Princeton University Press, 1988).
73. Neil Maher, *Nature's New Deal: The Civilian Conservation Corps and the Roots of the American Environmental Movement* (New York: Oxford University Press, 2007).
74. Gould, Pellow, and Schnaiberg, *Treadmill of Production.*
75. Aida Bey and Patrick Meyfroidt, "The Expansion of Large-Scale Tree Plantations: Detection, Pathways, and Development Trade-Offs," *Research Square* (2020), https://doi.org/10.21203/rs.3.rs-73265/v1.
76. Wolosin, *Large-Scale Forestation.*
77. FAO, *Global Forest Resources Assessment, 2020.*
78. Tom Brandeis et al., *Economic Dynamics of Forests and Forest Dynamics in the Southern United States* (Asheville, NC: Southern Research Station, United States Forest Service, 2017).
79. Rival and Levang, *Palms of Controversies.*
80. Rival and Levang, *Palms of Controversies.*
81. Arthur Mol and David Sonnenfeld, *Ecological Modernization Around the World: Perspectives and Critical Debates* (London: Frank Cass, 2000).
82. Norman Borlaug, "Feeding a Hungry World," *Science* 318 (2007): 359.
83. Ariel Lugo, "Comparison of Tropical Tree Plantations with Secondary Forests of a Similar Age," *Ecological Monographs* 62, no. 1(1992): 1–41.
84. Pedro Brancalion et al., "Exotic Eucalypts: From Demonized Trees to Allies of Tropical Forest Restoration," *Journal of Applied Ecology* 57 (2020): 55–66.
85. Lugo, "Understories."

86. FAO and UNEP, *The State of the World's Forests 2020: Forests, Biodiversity and People* (Rome: Food and Agricultural Organization of the United Nations, 2020).
87. Carlson et al., "Oil Palm Sustainability,"
88. E. Meijaard et al., "The Environmental Impacts of Palm Oil in Context," *Nature-Plants* 6 (2020): 1418–26.
89. Virginia Garcia et al., "Agricultural Intensification and Land Use Change: Assessing Country-level Induced Intensification, Land Sparing and Rebound Effect," *Environmental Research Letters* 15, no. 8 (2020): 085007.
90. Sloan, "Forest Transformation."

6. Agroforests I

1. A. Dobson et al., "Savannas Are Vital but Overlooked Carbon Sinks," *Science* 375 (2022): 6579.
2. J. F. Bastin et al., "The Global Tree Restoration Potential," *Science* 365 (2019): 76–79.
3. Niall Hanan and Michael Hill, *Savannas in a Changing Earth System: The NASA Terrestrial Ecology Project* (Washington, DC: National Aeronautics and Space Administration, 2012).
4. Douglas R. Shane, *Hoofprints in the Forest: Cattle Ranching and the Destruction of the Amazon Forest* (Philadelphia: Institute for the Study of Human Issues, 1986); Holly Gibbs et al., Tropical Forests Were the Primary Land Source for New Agricultural Lands in the 1980s and 1990s," *Proceedings of the National Academy of Science* 107, no. 38 (2010) 16732–37.
5. Robert Zomer et al., "Global Tree Cover and Biomass Carbon on Agricultural Land: The Contribution of Agroforestry to Global and National Carbon Budgets," *Science Reports* 6 (2016): 29987; Robert Zomer et al., "Trees on Farms: An Update and Reanalysis of Agroforestry's Global Extent and Socio-Ecological Characteristics" (Bogor, Indonesia: World Agroforestry Centre Southeast Asia Regional Program, 2014); Robert Zomer et al., "Trees on Farms: Analysis of Global Extent and Geographical Patterns of Agroforestry" (Nairobi: World Agroforestry Centre, 2009).
6. Zomer et al., "Trees on Farms."
7. Zomer et al., "Trees on Farms—Update."
8. Jeffrey Hoelle, *Rainforest Cowboys: The Rise of Ranching and Cattle Culture in Western Amazonia* (Austin: University of Texas Press, 2015); Shane, *Hoofprints*.
9. Thomas K. Rudel, "The Dynamics of Deforestation in the Wet and Dry Tropics: A Comparison with Policy Implications," *Forests* 8, no. 4 (2017): 108.
10. Thomas Rudel and Bruce Horowitz, *Tropical Deforestation: Small Farmers and Land Clearing in the Ecuadorian Amazon* (New York: Columbia University Press, 1993).
11. Amy Lerner et al., "The Spontaneous Emergence of Silvo-pastoral Landscapes in the Ecuadorian Amazon: Patterns and Processes," *Regional Environmental Change* 15, no. 7 (2015): 1421–31.

12. Diane Bates, "The Barbecho Crisis, La Plaga del Banco, and International Migration: Structural Adjustment in Ecuador's Southern Amazon," *Latin American Perspectives* 34 (2007): 108–22; Simon Maxwell, "Marginalized Peasants to the North of Santa Cruz: Avenues of Escape from the Barbecho Crisis," in *Land, People, and Planning in Contemporary Amazonia*, ed. F. Barbira-Scazziochio (Cambridge: Center for Latin American Studies, 1980), 162–70.
13. Lerner et al., "Spontaneous Emergence of Silvopastoral Landscapes."
14. International Union for the Conservation of Nature (IUCN), "Boosting Biodiversity in Colombia's Cattle and Coffee" (Gland, Switzerland: IUCN, 2022).
15. Roberto Porro, "Palms, Pastures, and Swidden Fields: The Grounded Political Ecology of Agro-Extractive/Shifting-Cultivator Peasants in Maranhao, Brazil," *Human Ecology* 33, no. 1 (2005): 17–56.
16. Peter May et al., "Babassu Palm in the Agroforestry Systems in Brazil's Mid-North Region," *Agroforestry Systems* 3, no. 3 (1985): 275–95.
17. Anthony Anderson, Peter May, and Michael Balick, *The Subsidy from Nature: Palm Forests, Peasantry, and Development on an Amazon Frontier* (New York: Columbia University Press, 1991).
18. Porro, "Palms, Pasture, and Swidden Fields."
19. Anderson, May, and Balick, *The Subsidy from Nature*.
20. Porro, "Palms, Pasture, and Swidden Fields."
21. Roberto Porro, "Land Use, Cattle Ranching, and the Concentration of Landownership in Maranhao, Brazil," in *Deforestation and Land Use in the Amazon*, ed. Charles H. Wood and Roberto Porro (Gainesville: University of Florida Press, 2002), 315–37; Porro, "Palms, Pasture, and Swidden Fields."
22. Megan McGroddy et al., "Effects of Pasture Management on Carbon Stocks: A Study from Four Communities in Southwestern Ecuador," *Biotropica* 47, no. 4 (2015): 407–15; Ziegler et al., "Carbon Outcomes."
23. J. M. Boffa, "Agroforestry Parklands in Sub-Saharan Africa" (Rome: Food and Agricultural Organization of the United Nations, 1999).
24. Peter Murphy and Ariel Lugo, "Ecology of Tropical Dry Forest," *Annual Review of Ecology and Systematics* 17 (1986): 67–88.
25. Boffa, "Agroforestry Parklands."
26. Emmanuel Chidumayo and Davison Gumbo, *The Dry Forests of Africa: Managing for Products and Services* (London: Earthscan, 2010).
27. Boffa, "Agroforestry Parklands."
28. Boffa, "Agroforestry Parklands."
29. Chris Reij, Gray Tappan, and Melinda Smale, "Agroenvironmental Transformation in the Sahel: Another Kind of 'Green Revolution'" (Washington, DC: International Food Policy Research Institute, 2009); Jan Sendzimir, Chris Reij, and Piotr Magnuszewski, "Rebuilding Resilience in the Sahel: Regreening in the Maradi and Zinder Regions of Niger," *Ecology and Society* 16, no. 3 (2011): 1.
30. Boffa, "Agroforestry Parklands."
31. Sendzimir, Reij, and Magnuszewski, "Rebuilding Resilience."
32. Chidumayo and Gumbo, *Dry Forests of Africa*.
33. Eric Haglund et al., "Dry Land Tree Management for Improved Household Livelihoods: Farmer Managed Natural Regeneration in Niger," *Journal of Environmental Management* 92 (2011): 1696–1705.

34. Boffa, "Agroforestry Parklands," 130.
35. Ruth Metzel and Florencia Montagnini, "From Farm to Forest: Factors Associated with Protecting and Planting Trees in a Panamanian Agricultural Landscape," *Bois et Forêts Tropiques* 322, no. 4 (2014): 3–15.
36. Jens Lund and Thorsten Treue, "Are We Getting There? Evidence of Decentralized Forest Management from the Tanzanian Miombo Woodlands," *World Development* 36, no. 12 (2008): 2780–800.
37. Boffa, "Agroforestry Parklands"; Reij, Tappan, and Smale, "Agroenvironmental Transformations in the Sahel."
38. Chidumayo and Gumbo, *Dry Forests of Africa.*
39. Reij, Tappan, and Smale, "Agroenvironmental Transformations in the Sahel."
40. Sendzimir, Reij, and Magnuszewski, "Rebuilding Resilience."
41. Boffa, "Agroforestry Parklands"; Reij, Tappan, and Smale, "Agroenvironmental Transformations in the Sahel."
42. Malavika Vyawahare, "Conflict and Climate Change Are Big Barriers for Africa's Great Green Wall," *Mongabay*, November 26, 2021.
43. David Kaimowitz, "The Meaning of Johannesburg," *Forest News*, CIFOR, September 19, 2002.

7. Agroforests II

1. Eduardo Brondizio, *The Amazonian Caboclo and the Açaí Palm: Forest Farmers in the Global Market* (New York: New York Botanical Garden, 2008); Genevieve Michon et al., "Domestic Forests: A New Paradigm for Integrating Local Communities' Forestry into Tropical Forest Science," *Ecology and Society* 12, no. 2 (2007): 1.
2. Meine Van Noordwijk, Ric Coe, and Fergus Sinclair, "Agroforestry Paradigms," in *Sustainable Development Through Trees on Farms: Agroforestry in Its Fifth Decade*, ed. M. van Noordwijk (Bogor, Indonesia: World Agroforestry [ICRAF] Southeast Asia Regional Program, 2019), 1–14.
3. Dennis Garrity, "Science Based Agroforestry and the Achievement of the Millennium Development Goals," in *World Agroforestry into the Future*, ed. D. Garrity et al. (Nairobi: World Agroforestry Centre, 2006), 3–9; Giovanni Ortolani, "Agroforestry: An Increasingly Popular Solution for a Hot, Hungry World," *Mongabay*, October 26, 2017.
4. Donald Kass, H. David Thurston, and Ken Schlather, "Sustainable Mulch-Based Cropping Systems with Trees," in *Agroforestry in Sustainable Agricultural Systems*, ed. L. Buck, J. Lassoie, and E. Fernandes (Boca Raton, FL: CRC Press, 1999), 367–85.
5. Gotz Schroth et al., "Conservation in Tropical Landscape Mosaics: The Case of the Cacao Landscape of Southern Bahia, Brazil," *Biodiversity and Conservation* 20 (2011): 1635–54.
6. Michon et al., "Domestic Forests."
7. Michon et al., "Domestic Forests."

8. M. Hua, E. Warren-Thomas, and T. Wanger, *Rubber Agroforestry: Feasibility at Scale* (Washington, DC: Mighty Earth, 2021).

9. Hubert Omont, Dominique Nicolas, and Diane Russell, "The Future of Perennial Tree Crops: What Role for Agroforestry?," in Garrity et al., *World Agroforestry into the Future*, 23–35.

10. Laurene Feintrenie and Patrice Levang, "Sumatra's Rubber Agroforests: Advent, Rise and Fall of a Sustainable Cropping System," *Small-scale Forestry* 8 (2009): 323–35.

11. Wil de Jong, "The Impact of Rubber on the Forest Landscape in Borneo," in *Agricultural Technologies and Tropical Deforestation*, ed. Arild Angelsen and David Kaimowitz (Wallingford, UK: CABI Publishing, 2001), 367–81.

12. Gede Wibawa, Sinung Hendratno, and Meine van Noordwijk, "Permanent Smallholder Rubber Agroforestry Systems in Sumatra, Indonesia," in *Slash and Burn Agriculture: The Search for Alternatives*, ed. C. Palm et al. (New York: Columbia University Press, 2005), 222–32.

13. Charlie Pye-Smith, "Rich Rewards for Rubber? Research in Indonesia is Exploring How Smallholders Can Increase Rubber Production, Retain Biodiversity and Provide Additional Environmental Benefits" (Nairobi: World Agroforestry Centre, 2011).

14. Pye-Smith, "Rich Rewards for Rubber."

15. Alan Ziegler et al., "Carbon Outcomes of Major Land-Cover Transitions in SE Asia: Great Uncertainties and REDD+ Policy Implications," *Global Change Biology* 18 (2012): 3087–99.

16. Thilde Bruun et al., "Environmental Consequences of the Demise in Swidden Cultivation in Southeast Asia: Carbon Storage and Soil Quality," *Human Ecology* 37 (2009): 375–88.

17. Alan Ziegler, Jefferson Fox, and Jiancho Xu, "The Rubber Juggernaut," *Science* 324 (2009): 1024–25.

18. Ziegler et al., "Carbon Outcomes."

19. Feintrenie and Levang, "Sumatra's Rubber Agroforests."

20. Pye-Smith, "Rich Rewards for Rubber."

21. Omont, Nicholas, and Russell, "The Future of Perennial Tree Crops."

22. Marius Wessel and P. Oquist-Wessel, "Cocoa Production in West Africa, a Review and Analysis of Recent Developments," *NJAS—Wageningen Journal of the Life Sciences* 74–75 (2015): 1–7.

23. Polly Hill, *The Migrant Cocoa-Farmers of Southern Ghana* (Cambridge: Cambridge University Press, 1963); Elsa Ordway, Greg Asner, and Eric Lambin, "Deforestation Risk Due to Commodity Crop Expansion in Sub-Saharan Africa," *Environmental Research Letters* 12 (2017): 044015.

24. C. Vancutsem et al., "Long-term (1990–2019) Monitoring of Forest Cover Changes in the Humid Tropics," *Science Advances* 7 (2021): eabe1603.

25. François Ruf, "The Myth of Complex Cocoa Agro-Forests," *Human Ecology* 39 (2011): 373–88; Wessel and Oquist-Wessel, "Cocoa Production in West Africa."

26. François Ruf, "Tree Crops as Deforestation and Reforestation Agents: The Case of Cocoa in Côte d'Ivoire and Sulawesi," in *Agricultural Technologies and Tropical Deforestation*, ed. A. Angelsen and D. Kaimowitz (Wallingford, UK: CABI Publishing, 2001), 291–315.

27. Ruf, "Tree Crops as Deforestation and Reforestation Agents."
28. Wiebke Niether et al., "Cocoa Agroforestry Systems Versus Monocultures: A Multi-Dimensional Meta-Analysis," *Environmental Research Letters* 15 (2020): 104085.
29. D. Sonwa et al., "Diversity of Plants in Cocoa Agroforests in the Humid Forest Zone of Southern Cameroon," *Biodiversity and Conservation* 16, no. 8 (2007): 2385–400.
30. Jim Gockowski and Denis Sonwa, "Cocoa Intensification Scenarios and Their Predicted Impact on CO2 Emissions, Biodiversity Conservation, and Rural Livelihoods in the Guinea Rain Forest of West Africa," *Environmental Management* 48 (2011): 307–21.
31. Niether et al., "Cocoa Agroforestry Systems Versus Monocultures."
32. Ruf, "The Myth of Complex Cocoa Agro-Forests."
33. Ruf, "The Myth of Complex Cocoa Agro-Forests."
34. Wessel and Oquist-Wessel, "Cocoa Production in West Africa."
35. Ruf, "Tree Crops as Deforestation and Reforestation Agents."
36. Ruf, "Tree Crops as Deforestation and Reforestation Agents."
37. Schroth, "Conservation in Tropical Landscape Mosaics."
38. Sean Kennedy, "The Power to Stay: Climate, Cocoa, and the Politics of Displacement," *Annals of the Association of American Geographers* 112, no. 2 (2022): 674–83.
39. International Union for the Conservation of Nature (IUCN), *Boosting Biodiversity in Colombia's Cattle and Coffee* (Gland, Switzerland: IUCN, 2022).
40. Harini Nagendra, Jane Southworth, and Catherine Tucker, "Accessibility as a Determinant of Landscape Transformation in Western Honduras: Linking Pattern and Process," *Landscape Ecology* 18 (2003): 141–58.
41. Mikaela Schmitt-Harsh et al., "Carbon Stocks in Coffee Agroforests and Mixed Dry Tropical Forests in the Western Highlands of Guatemala," *Agroforestry Systems* 86 (2012): 141–57.
42. Tanja Havemann et al., "Coffee in Dak Lak, Vietnam," in *Steps Toward Green: Policy Responses to the Environmental Footprint of Commodity Agriculture in East and Southeast Asia*, ed. S. Scherr et al. (Washington, DC: EcoAgriculture Partners and the World Bank, 2015), 99–122.
43. Virginia Garcia et al., "Agricultural Intensification and Land Use Change: Assessing Country-Level Induced Intensification, Land Sparing and Rebound Effect," *Environmental Research Letters* 15, no. 8 (2020): 085007.
44. Rhett Butler, "Commodity Eco-Certification Skyrockets, But Standards Slip," *Mongabay*, May 1, 2014.
45. X. Rueda and E. F. Lambin, "Responding to Globalization: Impacts of Certification on Colombian Small-scale Coffee Growers," *Ecology and Society* 18, no. 3 (2013): 21.
46. Ruth DeFries et al., "Is Voluntary Certification of Tropical Agricultural Commodities Achieving Sustainability Goals for Small-Scale Producers? A Review of the Evidence," *Environmental Research Letters* 12 (2017): 033001.
47. DeFries et al., "Voluntary Certification."
48. DeFries et al., "Voluntary Certification."
49. Michon, "Domestic Forests."

50. R. Netting, *Smallholders, Householders: Farm Families and the Ecology of Intensive, Sustainable Agriculture* (Stanford, CA: Stanford University Press, 1993).
51. Maia Call et al., "Socio-Environmental Drivers of Forest Change in Rural Uganda," *Land Use Policy* 62 (2017): 49–58.
52. Mary Tiffen, Michael Mortimore, and Frances Gichuki, *More People, Less Erosion: Environmental Recovery in Kenya* (Chichester, UK: Wiley, 1993).
53. Ruth Meinzen-Dick, "Women, Land, and Trees," in Garrity et al., *World Agroforestry into the Future*, 173–79.
54. Wangari Maathai, *The Green Belt Movement: Sharing the Approach and the Experience* (Herndon, VA: Lantern Books, 2004).
55. Robbie Corey-Boulet, "Despite Snags, Ethiopia Scales Up Massive Tree Planting Campaign," *Phys.org*, June 5, 2020, https://phys.org/news/2020-06-snags-ethiopia-scales-massive-tree-planting.html.
56. Lalisa Duguma et al., "From Tree Planting to Tree Growing: Rethinking Ecosystem Restoration Through Trees" (Nairobi: World Agroforestry, 2020).
57. Tiffen, Mortimore, and Gichuki, *More People, Less Erosion.*
58. Charlotte Boyd and Tom Slaymaker, *Re-examining the More People, Less Erosion Hypothesis: Special Case or Wider Trend?* (London: Overseas Development Institute, 2000).
59. Harold C. Conklin, "The Study of Shifting Cultivation," *Current Anthropology* 2, no. 1 (1961): 27–61.
60. Michon, "Domestic Forests."
61. Netting, *Smallholders, Householders.*
62. Netting, *Smallholders, Householders.*
63. Arya Dharmawan et al., "Dynamics of Rural Economy: A Socio-Economic Understanding of Oil Palm," *Land* 9, no. 7 (2020): 9070213.
64. Schelhas, John, "Land Use Choice and Change: Intensification and Diversification in the Lowland Tropics of Costa Rica," *Human Organization* 55, no. 3 (1996): 298–306.
65. Ruf, "Tree Crops as Deforestation and Reforestation Agents."

8. Resurgent Forests

1. John McNeill, *The Great Acceleration: An Environmental History of the Anthropocene Since 1945* (Cambridge, MA: Harvard University Press, 2016).
2. Miguel Pinedo-Vasquez et al., "Post-Boom Timber Production in Amazonia," *Human Ecology* 29 (2001): 219–39; Eduardo Brondizio, *The Amazonian Caboclo and the Açaí Palm: Forest Farmers in the Global Market* (New York: New York Botanical Garden, 2008).
3. Jeffrey Bentley, *Today There Is No Misery: The Ethnography of Farming in Northwest Portugal* (Tucson: University of Arizona Press, 1992).
4. Karl Polanyi, *The Great Transformation* (New York: Farrar and Rinehart, 1944).
5. Philippe Schmitter, "Still a Century of Corporatism?," *Review of Politics* 36, no. 1 (1974): 85–131; Oscar Molina and Martin Rhodes, "Corporatism: The

Past, Present, and Future of a Concept," *Annual Review of Political Science* 5 (2002): 305–31.

6. Charles Ragin, *The Comparative Method: Moving Beyond Qualitative and Quantitative Strategies* (Berkeley: University of California Press, 1987).

7. Sven Wunder et al., "REDD+ in Theory and Practice: How Local Projects Can Inform Jurisdictional Approaches," *Frontiers in Forests and Global Change* 3, no. 11 (2020). https://www.cifor.org/publications/pdf_files/articles/AWunder 2001.pdf.

8. Stibniati Atmadja, "Summary Analysis of REDD+ Projects, 2018–2020," v. 4.1, 2021, http://www.reddprojectsdatabase.org.

9. Amy Daniels, "Forest Expansion in Northwest Costa Rica: Conjuncture of the Global Market, Land Use Intensification, and Forest Protection," in *Reforesting Landscapes: Linking Pattern and Process*, ed. H. Nagendra and J. Southworth (Dordrecht: Springer, 2010), 227–52.

10. Chris Barrett et al., "Conserving Tropical Biodiversity Amid Weak Institutions," *Bioscience* 51, no. 6 (2001): 497–502.

11. L. Curran et al., "Lowland Forest Loss in Protected Areas of Indonesian Borneo," *Science* 303 (2004): 1000–1003.

12. Joel Migdal, *Strong Societies and Weak States: State-Society Relationships and State Capabilities in the Developing World* (Princeton, NJ: Princeton University Press, 1988).

13. David Kaimowitz, "The Meaning of Johannesburg," Forest News, CIFOR, September 19, 2002.

14. Max Boycoff, "Attitudes About Climate Change," presentation, Rutgers University, November 12, 2021.

15. The number links the description of a regional pattern of forestation to locations on the map in figure 8.1.

16. Chloe Ginsberg and Stephanie Keene, *At a Crossroads: Consequential Trends in Recognition of Community-Based Forest Tenure from 2002–2017* (Washington, DC: Rights and Resources Initiative, 2018); Frances Seymour and Jonas Busch, *Why Forests? Why Now? The Science, Economics, and Politics of Tropical Forests and Climate Change* (Washington, DC: Center for Global Development, 2016).

17. Atmadja, "Summary Analysis."

18. A. Sofia Nanni and H. Ricardo Grau, "Agricultural Adjustment, Population Dynamics and Forest Redistribution in a Subtropical Watershed of NW Argentina," *Regional Environmental Change* 14 (2014): 1641–49.

19. Amina Maharjan et al., "Understanding Rural Outmigration and Agricultural Land Use Change in the Gandaki Basin, Nepal," *Applied Geography* 124 (2020): e102278.

20. Jeffrey Bentley, "Bread Forests and New Fields: The Ecology of Reforestation and Forest Clearing Among Small-Woodland Owners in Portugal," *Journal of Forest History* 33, no. 4 (1989): 188–95.

21. Sarah Wilson et al., "Forest Ecosystem Transitions: The Ecological Dimensions of the Forest Transition," *Ecology and Society* 22, no. 4 (2017): 38.

22. Alain Rival and Patrice Levang, *Palms of Controversies: Oil Palm and Development Challenges* (Bogor, Indonesia: CIFOR, 2014).

23. John Zinda et al., "Dual Function Forests in the Returning Forests to Farmland Program and Flexibility in Environmental Policy in China," *Geoforum* 78 (2017): 119–32.

24. Atmadja, "Summary Analysis."

25. Tomas Sikor and Jacopo Baggio, "Can Smallholders Engage in Tree Plantations? An Entitlements Analysis from Vietnam," *World Development* 64 (2014): 101–12; David Kaczan, "Can Roads Contribute to Forest Transitions?"

26. Alan Ziegler, Jefferson Fox, and Jiancho Xu, "The Rubber Juggernaut," *Science* 324 (2009): 1024–25.

27. John Schelhas, Tom Brandeis, and Thomas K. Rudel, "Reforestation and Natural Regeneration in Forest Transitions: Patterns and Implications from the U.S. South," *Regional Environmental Change* 21, no. 1 (2021): 8.

28. M. Fagan et al., "The Expansion of Tree Plantations Across Tropical Biomes," *Nature Sustainability* 5 (2021): 681–88.

29. Greg Asner et al., "A Contemporary Assessment of Change in Humid Tropical Forests," *Conservation Biology* 23 (2009): 1386–95.

30. Amy Lerner et al., "The Spontaneous Emergence of Silvo-pastoral Landscapes in the Ecuadorian Amazon: Patterns and Processes," *Regional Environmental Change* 15, no. 7 (2015): 1421–31.

31. François Ruf, "Tree Crops as Deforestation and Reforestation Agents: The Case of Cocoa in Côte d'Ivoire and Sulawesi," in *Agricultural Technologies and Tropical Deforestation*, ed. A. Angelsen and D. Kaimowitz (Wallingford, UK: CABI Publishing, 2001), 291–315.

32. Marius Wessel and P. Oquist-Wessel, "Cocoa Production in West Africa, a Review and Analysis of Recent Developments," *NJAS—Wageningen Journal of the Life Sciences* 74–75 (2015): 1–7.

33. Ximena Rueda and Eric Lambin, "Responding to Globalization: Impacts of Certification on Colombian Small-scale Coffee Growers," *Ecology and Society* 18, no. 3 (2013): 21.

34. Ruth DeFries et al., "Is Voluntary Certification of Tropical Agricultural Commodities Achieving Sustainability Goals for Small-Scale Producers? A Review of the Evidence," *Environmental Research Letters* 12 (2017): 033001.

35. See https://www.greatgreenwall.org/about-great-green-wall.

36. Malakiva Vyawahare, "Conflict and Climate Change Are Big Barriers for Africa's Great Green Wall," *Mongabay*, November 26, 2021; Lalisa Duguma et al., "From Tree Planting to Tree Growing: Rethinking Ecosystem Restoration Through Trees" (Nairobi: World Agroforestry, 2020).

37. Wangari Maathai, *The Green Belt Movement: Sharing the Approach and the Experience* (Herndon, VA: Lantern Books, 2004).

38. Mark Poffenberger and Betsy McGean, *Village Voices, Forest Choices: Joint Forest Management in India* (Delhi: Oxford University Press, 1996).

39. David B. Bray, *Mexico's Community Forest Enterprises: Success on the Commons and the Seeds of a Good Anthropocene* (Tucson: University of Arizona Press, 2020).

40. Jens Lund and Thorsten Treue, "Are We Getting There? Evidence of Decentralized Forest Management from the Tanzanian Miombo Woodlands," *World Development* 36, no. 12 (2008): 2780–800.

41. Bray, *Mexico's Community Forest Enterprises*; Poffenberger and McGean, *Village Voices, Forest Choices.*
42. Elinor Ostrom, "Polycentric Systems for Coping with Collective Action and Global Environmental Change," *Global Environmental Change* 20 (2010): 550–57.

9. A Global Forest Transition?

1. Giorgos Kallis, "Radical Dematerialization and Degrowth," *Philosophical Transactions of the Royal Society A* 375, no. 2095 (2017): 0383; Diana Stuart, Ryan Gunderson, and Brian Petersen, *The Degrowth Alternative: A Path to Address our Environmental Crisis?* (New York: Routledge, 2021).
2. Mary Grigsby, *Buying Time and Getting By: The Voluntary Simplicity Movement* (Albany: SUNY Press, 2004).
3. John W. Reid and Thomas Lovejoy, *Ever Green: Saving Big Forests to Save the Planet* (New York: Norton, 2022).
4. Alexander Mather and Carolyn Needle, "The Forest Transition: A Theoretical Basis," *Area* 30, no. 2 (1998): 117–24; Alexander Mather and J. Fairbairn, "From Floods to Reforestation: The Forest Transition in Switzerland," *Environment and History* 6, no. 4 (2000): 399–421; Alexander Mather, Carolyn Needle, and James Coull, "From Resource Crisis to Sustainability: The Forest Transition in Denmark," *International Journal of Sustainable Development and World Ecology* 5 (1998): 182–93.
5. Thomas K. Rudel et al., "Whither the Forest Transition? Climate Change, Policy Responses, and Redistributed Forests in the Twenty-First Century," *Ambio* 49, no. 1 (2020): 74–84.
6. Alexander Mather, J. Fairbairn, and C. Needle, "The Course and Drivers of the Forest Transition: The Case of France," *Journal of Rural Studies* 15, no. 1 (1999): 65–90.
7. Alexander Gerschenkron, *Economic Backwardness in Historical Perspective* (Cambridge, MA: Harvard University Press, 1962).
8. Alexander Mather, "Recent Asian Transitions in Relation to Forest Transition Theory," *International Forestry Review* 9, no. 1 (2007): 491–502; Michael Wolosin, *Large-scale Forestation for Climate Mitigation: Lessons from South Korea, China, and India* (Washington, DC: Climate and Land Use Alliance, 2017).
9. Lewis et al., "Secondary Forests."
10. Norman Myers, *Conversion of Tropical Moist Forests* (Washington, DC: National Academies Press, 1980).
11. FAO, *Global Forest Resources Assessment, 2020* (Rome: Food and Agriculture Organization of the United Nations, 2020).
12. Rhett Butler, "Tropical Forests' Lost Decade: The 2010s," *Mongabay*, December 17, 2019.
13. Intergovernmental Science-Policy Platform on Biodiversity and Ecosystem Services (IPBES), *Nature's Dangerous Decline: Unprecedented Species Extinctions Accelerating* (New York: United Nations, 2021).

14. For the statistics, see https://aei.ag/2021/04/05/u-s-meat-consumption-trends-beef-pork-poultry-pandemic/Agricultural Economic Insights; Thomas K. Rudel, "The Variable Paths to Sustainable Intensification in Agriculture," *Regional Environmental Change* 20, no. 126 (2020): 126.

15. James Nations and Daniel Komer, "Rain Forests and the Hamburger Society," *Environment* 25, no. 3 (1983): 12–20.

16. Petterson Vale et al., "The Expansion of Intensive Beef Farming to the Brazilian Amazon," *Global Environmental Change* 57 (2019): 101922.

17. Vale, "Intensive Beef Farming."

18. Food and Agricultural Organization (FAO) and United Nations Environmental Program (UNEP), *The State of the World's Forests 2020: Forests, Biodiversity and People* (Rome: Food and Agricultural Organization of the United Nations, 2020).

19. John Meyer et al., "World Society and the Nation State," *American Journal of Sociology* 103, no. 1 (1997): 144–81.

20. Julia Flagg and Thomas K. Rudel, "Uneven Ambitions: Explaining National Differences in Proposed Emissions Reductions," *Human Ecology Review* 27, no. 1 (2021): 23–46.

21. FAO-UNEP, *Forests, Biodiversity and People.*

22. Frances Seymour and Jonas Busch, *Why Forests? Why Now? The Science, Economics, and Politics of Tropical Forests and Climate Change* (Washington, DC: Center for Global Development, 2016).

23. FAO-UNEP, *Forests, Biodiversity and People.*

24. R. Scott Frey, Thomas Dietz, and Linda Kalof, "Characteristics of Successful American Protest Groups," *American Journal of Sociology* 98, no. 2 (1992): 368–87; William Gamson, *The Strategy of Social Protest* (Homewood, IL: Dorsey Press, 1975).

25. Cultural Survival, "The Kayapo Bring their Case to the United States," March 1989. https://www.culturalsurvival.org/publications/cultural-survival-quarterly/kayapo-bring-their-case-united-states.

26. Thomas K. Rudel, "The Extractive Imperative in Populous Indigenous Territories: The Shuar, Copper Mining, and Environmental Injustices in the Ecuadorian Amazon," *Human Ecology* 46, no. 5 (2018): 717–24.

27. Peter Cronkleton et al., "Environmental Governance and the Emergence of Forest-Based Social Movements" (Bogor, Indonesia: Center for International Forestry Research, 2008).

28. Sven Wunder et al., "REDD+ in Theory and Practice: How Local Projects Can Inform Jurisdictional Approaches," *Frontiers in Forests and Global Change* 3, no. 11 (2020), https://www.cifor.org/publications/pdf_files/articles/AWunder2001.pdf.

29. Chris Barrett et al., "Conserving Tropical Biodiversity Amid Weak Institutions," *Bioscience* 51, no. 6 (2001): 497–502.

30. Yuta Masuda et al., "Emerging Research Needs and Policy Priorities for Advancing Land Tenure Security and Sustainable Development," in *Land Tenure Security and Sustainable Development*, ed. Margaret Holland, Yuta Masuda, and Brian Robinson (New York: Palgrave Macmillan, 2022), 313–26.

31. Basten Gokkon, "One Map to Rule Them All: Indonesia Launches Unified Land-Use Chart," *Mongabay*, December 13, 2018.

32. Max Bearak and Manuela Andreoni, "Brazil, Indonesia, and Congo Sign Rainforest Protection Pact," *New York Times*, November 14, 2022.
33. Virginia Garcia et al., "Agricultural Intensification and Land Use Change: Assessing Country-Level Induced Intensification, Land Sparing and Rebound Effect," *Environmental Research Letters* 15, no. 8 (2020): 085007.
34. Kimberly Carlson et al., "Effect of Oil Palm Sustainability Certification on Deforestation and Fire in Indonesia," *Proceedings of the National Academy of Sciences* 115, no. 1 (2018): 121–26.
35. Kingsley Davis, "The Theory of Change and Response in Modern Demographic History," *Population Index* 29, no. 4 (1963): 345–66.
36. Elinor Ostrom, "Polycentric Systems for Coping with Collective Action and Global Environmental Change," *Global Environmental Change* 20 (2010): 550–57.
37. David Kaimowitz, "The Meaning of Johannesburg," Forest News, CIFOR, September 19, 2002.
38. Ruth DeFries et al., "Planetary Opportunities: A Social Contract for Global Change Science to Contribute to a Sustainable Future," *Bioscience* 62, no. 6 (2012): 603–6.
39. Nancy Peluso, "Coercing Conservation: The Politics of State Resource Control," *Global Environmental Change* 3, no. 2 (1993): 199–217.

Bibliography

Aguirre Beltran, Gonzalo. *Regions of Refuge*. Washington, DC: Society of Applied Anthropology, 1979.

Albert, Federico. *Los bosques: Sus conservación, esplotación y fomento*. Santiago, Chile: Kosmos, 1913.

Alix-Garcia, Jennifer, et al. "Payments for Environmental Services Supported Social Capital While Increasing Land Management." *PNAS* 115 (2018): 7016–21.

Allen, C., A. Macalady, H. Chenchouni, D. Bachelet, N. McDowell, M. Vennetier, T. Kitzberger, et al. "A Global Overview of Drought and Heat-Induced Tree Mortality Reveals Emerging Climate Change Risks for Forests." *Forest Ecology and Management* 259, no. 4 (2010): 660–84.

Allen, Karen, and Steve Padgett Vasquez. "Forest Cover, Development, and Sustainability in Costa Rica: Can One Policy Fit All?" *Land Use Policy* 67 (2017): 212–21.

Alston, Lee, and Bernardo Mueller. "Legal Reserve Requirements in Brazilian Forests: Path Dependent Evolution of De Facto Legislation." *Revista Economia* 8, no. 4 (2007): 25–53.

Andela, N., D. Morton, L. Giglio, Y. Chen, G. van der Werf, P. Kasibhatla, R. DeFries, et al. "A Human-Driven Decline in Global Burned Area." *Science* 356 (2017): 1356–62.

Anderson, Anthony, Peter May, and Michael Balick. *The Subsidy from Nature: Palm Forests, Peasantry, and Development on an Amazon Frontier*. New York: Columbia University Press, 1991.

Andreoni, Manuela, Hiroko Tabuchi, and Albert Sun. "Destroying the Amazon for Leather Auto Seats." *New York Times*, November 18, 2021.

Angelsen, Arild. "REDD+ as Result-Based Aid: General Lessons and Bilateral Agreements of Norway." *Review of Development Economics* 21, no. 2 (2017): 237–64.

Angelsen, Arild, Maria Brockhaus, Amy Duchelle, Anne Larson, Christopher Martius, William Sunderlin, Louis Verchot, Grace Wong, and Sven Wunder. "Learning from REDD+: A Response to Fletcher et al." *Conservation Biology* 31, no. 3 (2017): 718–20.

Asner, Greg, Thomas K. Rudel, Ruth Defries, and T. Mitch Aide. "A Contemporary Assessment of Change in Humid Tropical Forests." *Conservation Biology* 23 (2009): 1386–95.

Assunção, J., C. Gandour, R. Rocha, and R. Rocha. "The Effect of Rural Credit on Deforestation: Evidence from the Brazilian Amazon." *The Economic Journal* 130, no. 626 (2020): 290–330.

Atmadja, Stibniati. "Summary Analysis of REDD+ Projects, 2018–2020, v. 4.1." 2021. http://www.reddprojectsdatabase.org.

Austin, Kemen, Mariano González-Roglich, Danica Schaffer-Smith, Amanda Schwantes, and Jennifer Swenson. "Trends in Size of Tropical Deforestation Events Signal Increasing Dominance of Industrial-Scale Drivers." *Environmental Research Letters* 5 (2017): 054009.

Baptista, Sandra, and Thomas K. Rudel. "Is the Atlantic Forest Re-emerging? Urbanization, Industrialization, and the Forest Transition in Santa Catarina, Southern Brazil." *Environmental Conservation* 33, no. 3 (2006): 195–202.

Barney, Keith. "China and the Production of Forest-Lands in Lao PDR: A Political Ecology of Transnational Enclosure." In *Taking Southeast Asia to Market: Commodities, Nature, and People in the Neoliberal Age*, ed. Joseph Nevins and Nancy Peluso, 91–107. Ithaca, NY: Cornell University Press, 2008.

Barr, Christopher, and Jeffrey Sayer. "The Political Economy of Reforestation and Forest Restoration in Asia-Pacific: Critical Issues for REDD+." *Biological Conservation* 154 (2012): 9–19.

Barrett, Chris, Katrina Brandon, Clark Gibson, and Heidi Gjertsen. "Conserving Tropical Biodiversity Amid Weak Institutions." *Bioscience* 51, no. 6 (2001): 497–502.

Barry, Deborah, and Ruth Meinzen-Dick. "The Invisible Map: Community Tenure Rights." In *The Social Lives of Forests: The Past, Present, and Future of Woodland Resurgence*, ed. Kathleen Morrison, Christine Padoch, and Susanna Hecht, 291–302. Chicago: University of Chicago Press, 2014.

Bastin, J. F., Y. Finegold, C. Garcia, D. Mollicone, M. Resende, et al. "The Global Tree Restoration Potential." *Science* 365 (2019): 76–79.

Bates, Diane. "The Barbecho Crisis, La Plaga del Banco, and International Migration: Structural Adjustment in Ecuador's Southern Amazon." *Latin American Perspectives* 34 (2007): 108–22.

Bates, Diane, and Thomas K. Rudel. "The Political Ecology of Tropical Rain Forest Conservation: A Cross-National Analysis." *Society and Natural Resources* 13, no. 7 (2000): 587–603.

Bearak, Max, and Manuela Andreoni. "Brazil, Indonesia, and Congo Sign Rainforest Protection Pact." *New York Times*, November 14, 2022.

Beck, Philip, and Milton Forster. *Six Rural Problem Areas: Relief, Resources, and Rehabilitation*. Washington, DC: Works Progress Administration, Division of Social Research, 1935.

Bentley, Jeffrey. "Bread Forests and New Fields: The Ecology of Reforestation and Forest Clearing Among Small-Woodland Owners in Portugal." *Journal of Forest History* 33, no. 4 (1989): 188–95.

——. *Today There Is No Misery: The Ethnography of Farming in Northwest Portugal*. Tucson: University of Arizona Press, 1992.

Bernal, B., L. Murray, and T. Pearson. "Carbon Dioxide Removal Rates from Forest Landscape Restoration Activities." *Carbon Balance Management* 13 (2018): 22.

Bey, Aida, and Patrick Meyfroidt. "The Expansion of Large-Scale Tree Plantations: Detection, Pathways, and Development Trade-Offs." Research Square. 2020. https://doi.org/10.21203/rs.3.rs-73265/v1.

Boffa, J. M. "Agroforestry Parklands in Sub-Saharan Africa." Rome: Food and Agricultural Organization of the United Nations, 1999.

Borlaug, Norman. "Feeding a Hungry World." *Science* 318 (2007): 359.

Borras, Saturnino, Ruth Hall, Ian Scoones, Ben White, and Wendy Wolford. "Towards a Better Understanding of Global Land Grabbing: An Editorial Introduction." *Journal of Peasant Studies* 38, no. 2 (2011): 209–16.

Boserup, Ester. *The Conditions of Agricultural Growth: The Economics of Agrarian Change under Population Pressure.* London: Allen and Unwin, 1965.

Boucher, Doug, Sarah Roquemore, and Estrellita Fitzhugh. "Brazil's Success in Reducing Deforestation." *Tropical Conservation Science* 6, no. 3 (2013): 426–45.

Boycoff, Max. "Attitudes About Climate Change." Presentation, Rutgers University, November 12, 2021.

Boyd, Charlotte, and Tom Slaymaker. *Re-examining the More People, Less Erosion Hypothesis: Special Case or Wider Trend?* London: Overseas Development Institute, 2000.

Brancalion, Pedro, Nino Amazonas, Robin Chazdon, Juliano van Melis, Ricardo Rodrigues, Carina Silva, Taisi Sorrini, and Karen Holl. "Exotic Eucalypts: From Demonized Trees to Allies of Tropical Forest Restoration." *Journal of Applied Ecology* 57 (2020): 55–66.

Brandeis, Tom, Andrew Hartsell, James Bentley, and Consuelo Brandeis. *Economic Dynamics of Forests and Forest Dynamics in the Southern United States.* Asheville, NC: Southern Research Station, United States Forest Service, 2017.

Bray, David B. *Mexico's Community Forest Enterprises: Success on the Commons and the Seeds of a Good Anthropocene.* Tucson: University of Arizona Press, 2020.

Brechin, Steven, and William Fenner. "Karl Polanyi's Environmental Sociology: A Primer." *Environmental Sociology* 3, no. 3 (2017): 1–10.

Brondizio, Eduardo. *The Amazonian Caboclo and the Açaí Palm: Forest Farmers in the Global Market.* New York: New York Botanical Garden, 2008.

Bruun, Thilde, Andreas de Neergaard, Deborah Lawrence, and Alan Ziegler. "Environmental Consequences of the Demise in Swidden Cultivation in Southeast Asia: Carbon Storage and Soil Quality." *Human Ecology* 37 (2009): 375–88.

Butler, Rhett. "Commodity Eco-Certification Skyrockets, but Standards Slip." *Mongabay*, May 1, 2014.

——. "Tropical Forests' Lost Decade: The 2010s." *Mongabay*, December 17, 2019.

Call, Maia, Tony Mayer, Samuel Sellers, Diamond Ebanks, Margit Bertalin, Elizabeth Nebie, and Clark Gray. "Socio-Environmental Drivers of Forest Change in Rural Uganda." *Land Use Policy* 62 (2017): 49–58.

Carlson, Kimberly, Robert Heilmayr, Holly Gibbs, Praveen Noojipady, David Burns, Douglas Morton, Nathalie Walker, Gary Paoli, and Claire Kremen. "Effect of Oil Palm Sustainability Certification on Deforestation and Fire in Indonesia." *Proceedings of the National Academy of Sciences* 115, no. 1 (2018): 121–26.

Cassano, Camilla, Gotz Schroth, Deborah Faria, and Jacques Delable. "Conservation in Tropical Landscape Mosaics: The Case of the Cacao Landscape of Southern Bahia, Brazil." *Biodiversity and Conservation* 20, no. 8 (2011): 1635–54.

Chan, Nyein, and Shinya Takeda. "The Transition Away from Swidden Agriculture and Trends in Biomass Accumulation in Fallow Forests: Case Studies in the Southern Chin Hills of Myanmar." *Mountain Research and Development* 36, no. 3 (2016): 320–31.

Charruadas, Paulo, and Chloe Deligne. "Cities Hiding the Forests: Wood Supply, Hinterlands, and Urban Agency in the Southern Low Countries, Thirteen—Eighteenth Centuries." In *Urbanizing Nature: Actors and Agency (Dis)Connecting Cities and Nature Since 1500*, ed. T. Soens, D. Schott, M. Toyka-Seid, and B. De Munck, 367–85. London: Taylor and Francis, 2019.

Chazdon, Ruth. *Second Growth: The Promise of Tropical Forest Regeneration in an Age of Deforestation*. Chicago: University of Chicago Press, 2014.

Chazdon, Ruth, Pedro Brancalion, Lars Laestadius, Aoife Bennett-Curry, Kathleen Buckingham, Chetar Kumar, Julian Moll-Rocek, Ima Vieira, and Sarah Wilson. "When Is a Forest a Forest? Forest Concepts and Definitions in the Era of Forest and Landscape Restoration." *Ambio* 45 (2016): 538–50.

Chen, C., T. Park, X. Wang, S. Piao, B. Xu, R. Chaturvedi, R. Fuchs, et al. "China and India Lead in Greening of the World Through Land-use Management." *Nature Sustainability* 2 (2019): 122–29.

Chen, Jai, John Zinda, and E. Ting Yeh. "Recasting the Rural: State, Society, and Environment in Contemporary China." *Geoforum* 78 (2017): 83–88.

Chen, Jing, Weimin Ju, Phillippe Ciais, Nicholas Viovy, Ronghao Liu, Yang Liu, and Xuehe Lu. "Vegetation Structural Change Since 1981 Significantly Enhanced the Terrestrial Carbon Sink." *Nature Communications* 10 (2019): 4259.

Chidumayo, Emmanuel, and Davison Gumbo. *The Dry Forests of Africa: Managing for Products and Services*. London: Earthscan, 2010.

Clark, William, and Alicia Harley. "Sustainability Science: Towards a Synthesis." *Annual Review of Environment and Resources* 45 (2020): 331–86.

Collins, S., S. Carpenter, S. Swinton, D. Orenstein, D. Childers, T. Gragson, N. Grimm, et al. "An Integrated Conceptual Framework for Long-Term Social–Ecological Research." *Frontiers in Ecology and the Environment* 9 (2011): 351–57.

Conklin, Harold C. "The Study of Shifting Cultivation." *Current Anthropology* 2, no. 1 (1961): 27–61.

Cook-Patton, S., S. Leavitt, D. Gibbs, N. Harris, K. Lister, K. Anderson-Teixeira, R. Briggs, et al. "Mapping Carbon Accumulation Potential from Global Natural Forest Regrowth." *Nature* 585 (2020): 545–50.

Corey-Boulet, Robbie. "Despite Snags, Ethiopia Scales Up Massive Tree Planting Campaign." Phys.org. June 5, 2020. https://phys.org/news/2020-06-snags-ethiopia-scales-massive-tree-planting.html.

Cronkleton, Peter, Peter Taylor, Deborah Barry, Samantha Stone-Jovicich, and Marianne Schmink. "Environmental Governance and the Emergence of Forest-Based Social Movements." Bogor, Indonesia: Center for International Forestry Research, 2008.

Cultural Survival. "The Kayapo Bring Their Case to the United States." 1989. https://www.culturalsurvival.org/publications/cultural-survival-quarterly/kayapo-bring-their-case-united-states.

Curran, L., N. Trigg, A. K. McDonald, D. Astiani, Y. M. Hardiono, P. Siregar, I. Caniago, and E. Kasischke. "Lowland Forest Loss in Protected Areas of Indonesian Borneo." *Science* 303 (2004): 1000–1003.

Curtis, Philip, Christy Slay, Nancy Harris, Alexandra Tyukavina, and Matthew Hansen. "Classifying Drivers of Global Forest Loss." *Science* 361 (2018): 1108–11.

Dahl, Robert. *Who Governs? Democracy and Power in an American City.* New Haven, CT: Yale University Press, 1961.

Dai, Aiguo. "Drought Under Global Warming: A Review." *WIRES—Climate Change* 2 (2011): 45.

——. "Increasing Drought Under Global Warming in Observations and Models." *Nature Climate Change* 3 (2013): 52–58.

Daniel, Carrie, Amanda Cross, Corinna Koebnick, and Rashmi Sinha. "Trends in Meat Consumption in the U.S." *Public Health Nutrition.* 14 (2011): 575–83.

Daniels, Amy. "Forest Expansion in Northwest Costa Rica: Conjuncture of the Global Market, Land Use Intensification, and Forest Protection." In *Reforesting Landscapes: Linking Pattern and Process*, ed. H. Nagendra and J. Southworth, 227–52. Dordrecht: Springer, 2010.

D'Antona, Alvaro, Leah Van Wey, and Corey Hayashi. "Property Size and Land Cover Change in the Brazilian Amazon." *Population and Environment* 27 (2006): 373–96.

Dauvergne, Peter. *Shadows in the Forest: Japan and the Politics of Timber in Southeast Asia.* Cambridge, MA: MIT Press, 1997.

Dave, R., C. Saint-Laurent, L. Murray, G. Antunes Daldegan, R. Brouwer, R. C. de Mattos Scaramuzza, et al. *Second Bonn Challenge Progress Report: Application of the Barometer in 2018.* Gland, Switzerland: IUCN, 2019.

Davis, Kingsley. "The Theory of Change and Response in Modern Demographic History." *Population Index* 29, no. 4 (1963): 345–66.

Davis, Kyle, Kailiang Yu, Maria Rulli, Lonn Pichdara, and Paolo D'Odorico. "Accelerated Deforestation Driven by Large-Scale Land Acquisitions in Cambodia." *Nature Geosciences* 8, no. 10 (2015): 772–75.

DeFries, Ruth, Erle Ellis, Pamela Matson, F. Chapin, Arun Agrawal, Billie L. Turner, Paul Crutzen, et al. "Planetary Opportunities: A Social Contract for Global Change Science to Contribute to a Sustainable Future." *Bioscience* 62, no. 6 (2012): 603–6.

DeFries, Ruth, Jessica Fanzo, Pinki Mondal, Rosalie Remans, and Steven Wood. "Is Voluntary Certification of Tropical Agricultural Commodities Achieving Sustainability Goals for Small-Scale Producers? A Review of the Evidence." *Environmental Research Letters* 12 (2017): 033001.

DeFries, Ruth, Thomas K. Rudel, Maria Uriarte, and Matthew Hansen. "Deforestation Driven by Urban Population Growth and Agricultural Trade in the Twenty-First Century." *Nature Geosciences* 3 (2010): 178–81.

de Jong, Wil. "The Impact of Rubber on the Forest Landscape in Borneo." In *Agricultural Technologies and Tropical Deforestation*, ed. Arild Angelsen and David Kaimowitz, 367–81. Wallingford, UK: CABI Publishing, 2001.

Descals, Adria, David Gaveau, Aleixandre Verger, Douglas Sheil, Daisuke Naito, and Josep Penuelas. "Unprecedented Fire Activity Above the Arctic Circle Linked to Rising Temperatures." *Science* 378, no. 6619 (2022): 532–37.

Dharmawan, Arya, Dayh Mardiyaningsih, Heru Komarudin, Jaboury Ghazoul, Pablo Pacheco, and Faris Rahmadian. "Dynamics of Rural Economy: A Socio-Economic Understanding of Oil Palm." *Land* 9, no. 7 (2020): 9070213.

Dinerstein, E., D. Olson, A. Joshi, C. Vynne, N. Burgess, E. Wikramanayake, N. Hahn, et al. "An Ecoregion-Based Approach to Protecting Half the Terrestrial Realm." *Bioscience* 67 (2017): 534–45.

Directorate General of Estate Crops. "Tree Crop Estate Statistics of Indonesia, Palm Oil, 2015–2017." Jakarta, Indonesia: Department of Agriculture, 2016.

Dobson, A., G. Hopcraft, S. Mduma, J. Ogutu, J. Fryxell, T. Anderson, S. Archibaul, et al. "Savannas Are Vital but Overlooked Carbon Sinks." *Science* 375 (2022): 6579.

Drummond, Mark, and Thomas Loveland. "Land Use Pressure and a Transition to Forest-Cover Loss in the Eastern United States." *BioScience* 60, no. 4 (2010): 286–98.

Dryzek, John, Christian Hunold, David Schlosberg, David Downes, and Hans Kristian Hernes. "The Environmental Transformation of the State: The USA, Norway, Germany, and the UK." *Political Studies* 50 (2002): 659–82.

Duguma, Lalisa, Peter Minang, Ermias Aynekulu, Sammy Carsan, Judith Nzyoka, Alagie Bah, and Ramni Jamnadass. "From Tree Planting to Tree Growing: Rethinking Ecosystem Restoration Through Trees." Nairobi: World Agroforestry, 2020.

Duram, Leslie, Jon Bathgate, and Christina Ray. "A Local Example of Land Use Change: Southern Illinois, 1807, 1938, 1993." *Professional Geographer* 56, no. 1 (2004): 127–40.

Duran, America, and Olga Barbosa. "Seeing Chile's Forest for the Tree Plantations." *Science* 365, no. 6460 (2019): 1388.

Durkheim, Émile. *The Division of Labour in Society.* 1897. Glencoe, IL: Free Press, 1964.

Edelman, Marc. "Extensive Land Use and the Logic of the Latifundio. A Case Study in Guanacaste Province, Costa Rica." *Human Ecology* 13, no. 2 (1985): 153–85.

Elias, F., J. Ferreira, G. Lennox, S. Berenguer, S. Ferreira, G. Schwartz, L. Oliveira Melo, D. Reis, et al. "Assessing the Growth and Climate Sensitivity of Secondary Forests in Climate Sensitive Amazonian Landscapes." *Ecology* 101, no. 3 (2020): ecy2954.

Estel, Stephan, et al. "Mapping Farmland Abandonment and Recultivation Across Europe Using MODIS NDVI Time Series." *Remote Sensing of the Environment* 163 (2015): 312–25.

Fa, J., J. Watson, I. Leiper, P. Potapov, T. Evans, N. Burgess, Z. Molnár, et al. "Importance of Indigenous Peoples' Lands for the Conservation of Intact Forest Landscapes." *Frontiers in Ecology and the Environment* 18, no. 3 (2020): 135–40.

Fagan, M. E., D. H. Kim, W. Settle, L. Ferry, J. Drew, H. Carlson, J. Slaughter, et al. "The Expansion of Tree Plantations Across Tropical Biomes." *Nature Sustainability* 5 (2021): 681–88.

Feder, Ernesto. *The Rape of the Peasantry: Latin America's Landholding System.* New York: Anchor Books, 1971.

Feintrenie, Laurene, and Patrice Levang. "Sumatra's Rubber Agroforests: Advent, Rise and Fall of a Sustainable Cropping System." *Small-Scale Forestry* 8 (2009): 323–35.

Fite, Gilbert C. *Cotton Fields No More: Southern Agriculture, 1865–1980*. Lexington: University of Kentucky Press, 1983.

Flagg, Julia, and Thomas K. Rudel. "Uneven Ambitions: Explaining National Differences in Proposed Emissions Reductions." *Human Ecology Review* 27, no. 1(2021): 23–46.

Food and Agricultural Organization (FAO). *Global Forest Resources Assessment, 2010*. Rome: Food and Agriculture Organization of the United Nations, 2010.

——. *Global Forest Resources Assessment, 2020*. Rome: Food and Agriculture Organization of the United Nations, 2020.

——. *Mountains as the Water Towers of the World: A Call for Action on the Sustainable Development Goals (SDGs)*. Rome: Food and Agricultural Organization of the United Nations, 2014.

——. *The State of the World's Land and Water Resources for Food and Agriculture (SOLAW)—Managing Systems at Risk*. Rome: Food and Agriculture Organization of the United Nations, 2011.

Food and Agricultural Organization (FAO) and United Nations Environmental Program (UNEP). *The State of the World's Forests 2020: Forests, Biodiversity and People*. Rome: Food and Agricultural Organization of the United Nations, 2020.

Foresta, Ronald. *Amazon Conservation During the Age of Development*. Gainesville: University of Florida Press, 1991.

Foster, David. *Wildlands and Woodlands: Farmlands and Communities—Broadening the Vision for New England*. Petersham, MA: Harvard Forest, 2017.

Foster, David, and John Aber. *Forests in Time: The Environmental Consequences of 1000 Years of Change in New England*. New Haven, CT: Yale University Press, 2004.

Frey, R. Scott, Thomas Dietz, and Linda Kalof. "Characteristics of Successful American Protest Groups." *American Journal of Sociology* 98, no. 2 (1992): 368–87.

Gamson, William. *The Strategy of Social Protest*. Homewood, IL: Dorsey Press, 1975.

Garcia, C., S. Savilaakso, R. Verburg, V. Gutierrez, S. Wilson, C. Krug, M. Sassen, et al. "The Global Forest Transition as a Human Affair." *One Earth* 2 (2020): 417–28.

Garcia, Virginia, Frederic Gaspart, Thomas Kastner, and Patrick Meyfroidt. "Agricultural Intensification and Land Use Change: Assessing Country-Level Induced Intensification, Land Sparing and Rebound Effect." *Environmental Research Letters* 15, no. 8 (2020): 085007.

Garnett, S., D. Burgess, J. Fa, A. Fernández-llamazares, Z. Molnar, C. Robinson, J. Watson, et al. "A Spatial Overview of the Global Importance of Indigenous Lands for Conservation." *Nature Sustainability* 1 (2018): 369–74.

Garrity, Dennis. 2006. "Science Based Agroforestry and the Achievement of the Millennium Development Goals." In *World Agroforestry into the Future*, ed. D. Garrity, A. Okono, M. Grayson, and S. Parrott, 3–9. Nairobi: World Agroforestry Centre, 2006.

Gatti, L., L. Basso, J. Miller, M. Gloor, L. Gatti Domingues, H. Cassol, R. Neves, et al. "Amazonia as a Carbon Source Linked to Deforestation and Climate Change." *Nature* 595, no. 7867 (2021): 388–93.

Gentry, Alwyn. "Patterns of Diversity and Floristic Composition in Neotropical Montane Forests." In *Biodiversity and Conservation of Neo-tropical Montane Forests*, ed. S. Churchill, 103–26. New York: New York Botanical Garden, 1995.

George, Andrew, and Alexander Bennett. *Case Studies and Theory Development in the Social Sciences.* Cambridge, MA: MIT Press, 2005.

Gerschenkron, Alexander. *Economic Backwardness in Historical Perspective.* Cambridge, MA: Harvard University Press, 1962.

Gibbs, Holly, Aaron Ruesch, Jonathan Foley, Navin Ramankutty, Frederic Achard, and Peter Holmgren. "Tropical Forests Were the Primary Land Source for New Agricultural Lands in the 1980s and 1990s." *Proceedings of the National Academy of Science* 107, no. 38 (2010) 16732–37.

Ginsberg, Chloe, and Stephanie Keene. *At a Crossroads: Consequential Trends in Recognition of Community-Based Forest Tenure from 2002–2017.* Washington, DC: Rights and Resources Initiative, 2018.

Goldthwait, James. "A Town That Has Gone Downhill." *Geographical Review* 17, no. 4 (1927): 527–52.

Gockowski, Jim, and Denis Sonwa. "Cocoa Intensification Scenarios and Their Predicted Impact on CO_2 Emissions, Biodiversity Conservation, and Rural Livelihoods in the Guinea Rain Forest of West Africa." *Environmental Management* 48 (2011): 307–21.

Gokkon, Basten. "One Map to Rule Them All: Indonesia Launches Unified Land-Use Chart." *Mongabay,* December 13, 2018.

Goodwin, Geoff. "Rethinking the Double Movement: Expanding the Frontiers of Polanyian Analysis in the Global South." *Development and Change* 49, no. 5 (2018): 1268–90.

Gould, Kenneth, David Pellow, and Allan Schnaiberg. *Treadmill of Production: Injustice and Unsustainability in the Global Economy.* Boulder, CO: Paradigm Publishers, 2008.

Grigsby, Mary. *Buying Time and Getting By: The Voluntary Simplicity Movement.* Albany: SUNY Press, 2004.

Griscom, B., J. Adams, P. Ellis, R. Houghton, G. Lomax, et al. "Natural Climate Solutions." *Proceedings of the National Academy of Sciences* 144 (2017): 11645–50.

Gu, Huan, Christopher Williams, Natalia Hasler, and Yu Zhou. "The Carbon Balance of the Southeastern U.S. Forest Sector as Driven by Recent Disturbance Trends." *Journal of Geophysical Research: Biogeosciences* 124 (2019): 2786–803.

Haglund, Eric, Jupiter Ndjeunga, Laura Snook, and Dov Pasternak. "Dry Land Tree Management for Improved Household Livelihoods: Farmer Managed Natural Regeneration in Niger." *Journal of Environmental Management* 92 (2011): 1696–1705.

Haig, Irvine, L. V. Teesdale, Philip Briegleb, Burnett Payne, and Martin Haertel. *Forest Resources of Chile as a Basis for Industrial Expansion.* Washington, DC: Forest Service, U.S. Department of Agriculture, 1946.

Hanan, Niall, and Michael Hill. *Savannas in a Changing Earth System. The NASA Terrestrial Ecology Project.* Washington, DC: National Aeronautics and Space Administration, 2012.

Havemann, Tanja, Samiksha Nair, Emilie Cassou, and Steven Jaffee. "Coffee in Dak Lak, Vietnam." In *Steps Toward Green: Policy Responses to the Environmental Footprint of Commodity Agriculture in East and Southeast Asia,* ed. S. Scherr, K. Mankad, S. Jaffee, and C. Negra, 99–122. Washington, DC: EcoAgriculture Partners and the World Bank, 2015.

Hecht, Susanna, Anastasia Yang, Bimbaka Basnett, Christine Padoch, and Nancy Peluso. "People in Motion, Forests in Transition: Trends in Migration, Urbanization, and Remittances and Their Effects on Tropical Forests." Bogor, Indonesia: CIFOR, 2015.

Heilmayr, Robert, Cristian Echeverria, R. Fuentes, and Eric Lambin. "A Plantation-Dominated Forest Transition in Chile." *Applied Geography* 75 (2016): 71–82.

Heilmayr, Robert, Cristian Echeverria, and Eric Lambin. "Impacts of Chilean Forest Subsidies on Forest Cover, Carbon and Biodiversity." *Nature—Sustainability* 3 (2020): 701–9.

Heilmayr, Robert, and Eric Lambin. "Impacts of Nonstate, Market-Driven Governance on Chilean Forests." *Proceedings of the Natural Academy of Sciences* 113, no. 11 (2016): 2910–15.

Hertel, Thomas. "Growing Food and Protecting Nature Don't Have to Conflict—Here's How They Can Work Together." *The Conversation.* March 9, 2021. https://theconversation.com/growing-food-and-protecting-nature-dont-have-to-conflict-heres-how-they-can-work-together-146069.

Hertel, Thomas, Navin Ramankutty, and Uris Baldos. "Global Market Integration Increases Likelihood that a Future African Green Revolution Could Increase Cropland Use and CO_2 Emissions." *Proceedings of the National Academy of Sciences* 111, no. 38 (2014): 13799–804.

Hill, Polly. *The Migrant Cocoa-Farmers of Southern Ghana.* Cambridge: Cambridge University Press, 1963.

Hoelle, Jeffrey. *Rainforest Cowboys: The Rise of Ranching and Cattle Culture in Western Amazonia.* Austin: University of Texas Press, 2015.

Hong, C., H. Zhao, Y. Qin, J. Burney, J. Pongratz, K. Hartung, Y. Liu, et al. "Land Use Emissions Embodied in International Trade." *Science* 376, no. 6593 (2022): 597–603.

Hua, M., E. Warren-Thomas, and T. Wanger. *Rubber Agroforestry: Feasibility at Scale.* Washington, DC: Mighty Earth, 2021.

Igarape Institute. *Connecting the Dots: Territories and Trajectories of Environmental Crime in the Brazilian Amazon and Beyond.* Rio de Janeiro: Igarape Institute, 2022.

Instituto Forestal (INFOR). *Estadísticas Forestales.* Santiago, Chile: INFOR, 1986.

Intergovernmental Panel on Climate Change (IPCC). *AR6 Climate Change: The Physical Basis.* Geneva: IPCC Secretariat, 2021.

Intergovernmental Science-Policy Platform on Biodiversity and Ecosystem Services (IPBES) *Nature's Dangerous Decline: Unprecedented Species Extinctions Accelerating.* New York: United Nations, 2021.

International Centre for Research in Agroforestry (ICRAF). "Annual Report 1993." Nairobi: International Centre for Research in Agroforestry, 1993.

International Union for the Conservation of Nature (IUCN). "Boosting Biodiversity in Colombia's Cattle and Coffee." Gland, Switzerland: IUCN, 2022.

Jadin, Isaline, Patrick Meyfroidt, and Eric Lambin. "International Trade, and Land Use Intensification and Spatial Reorganization Explain Costa Rica's Forest Transition." *Environmental Research Letters* 11 (2016): 035005.

Jayne, T. S., Jordan Chamberlin, and Derek Headey. "Land Pressures, the Evolution of Farming Systems, and Development Strategies in Africa: A Synthesis." *Food Policy* 48 (2014): 1–17.

Jelsma, Idsert, Maja Slingerland, Ken Giller, and Jos Bijman. "Collective Action in a Smallholder Oil Palm Production System in Indonesia: The Key to Sustainable and Inclusive Smallholder Oil Palm." *Journal of Rural Studies* 54 (2017): 198–210.

Jong, Hans Nicholas. "Indonesia Forest Clearing Ban Is Made Permanent, but Labelled Propaganda." *Mongabay*, August 14, 2019.

——. "Indonesia Terminates Agreement with Norway on $1b REDD+ scheme." *Mongabay*, September 10, 2021.

——. "RSPO Fails to Deliver on Environmental and Social Sustainability, Study Finds." *Mongabay*, July 11, 2018.

Kaczan, David. "Can Roads Contribute to Forest Transitions?" *World Development* 129 (2020): 104898.

Kaimowitz, David. "The Meaning of Johannesburg." *Forest News*, CIFOR, September 19, 2002.

Kallis, Giorgos. "Radical Dematerialization and Degrowth." *Philosophical Transactions of the Royal Society A* 375, no. 2095 (2017): 0383.

Kass, Donald, H. David Thurston, and Ken Schlather. "Sustainable Mulch-Based Cropping Systems with Trees." *Agroforestry in Sustainable Agricultural Systems*, ed. L. Buck, J. Lassoie, and E. Fernandes,. Boca Raton, FL: CRC Press, 1999.

Kennedy, Sean. "The Power to Stay: Climate, Cocoa, and the Politics of Displacement." *Annals of the Association of American Geographers* 112, no. 2 (2022): 674–83.

Kohler, Florent, and Eduardo Brondizio. "Considering the Needs of Indigenous and Local Populations in Conservation Programs." *Conservation Biology* 31 (2017): 245–51.

Kuemmerle, Tobias, et al. "Hotspots of Land Use Change in Europe." *Environmental Research Letters* 11 (2016): 064020.

Kull, Christian. "Forest Transitions: A New Conceptual Scheme." *Geografica Helvetica* 72 (2017): 465–74.

Laidlaw, Zoe, and Alan Lester. *Indigenous Communities and Settler Colonialism: Landholding, Loss, and Survival in an Interconnected World.* New York: Palgrave MacMillan, 2015.

Lara, Antonio, María Eugenia Solari, María Del Rosario Prieto, and María Paz Peña. "Reconstrucción de la cobertura de la vegetación y uso del suelo hacia 1550 y sus cambios a 2007 en la ecorregión de los bosques valdivianos lluviosos de Chile (35°–43°30 S)." *Bosque (Valdivia)* 33 (2012): 13–23.

Larson, Anne M., and Ganga Dahal. "Forest Tenure Reform: New Resource Rights for Forest Based Communities." *Conservation and Society* 10, no. 2 (2012): 77–90.

Lerner, Amy, Thomas K. Rudel, Laura Schneider, Megan McGroddy, Diana Burbano, and Carlos Mena. "The Spontaneous Emergence of Silvo-pastoral Landscapes in the Ecuadorian Amazon: Patterns and Processes." *Regional Environmental Change* 15, no. 7 (2015): 1421–31.

Lestrelin, Guillame, Jean Cristophe Castella, and Jeremy Bourgoin. "Territorializing Sustainable Development: The Politics of Land-Use Planning in Laos." *Journal of Contemporary Asia.* (2012): 1–22.

Lewis, Simon L., Charlotte Wheeler, Edward Mitchard, and Alexander Koch. "Regenerate Natural Forests to Store Carbon." *Nature* 568 (2019): 25–28.

Leyva Galan, Angel. "Metodología para el desarollo de la biodiversidad vegetal." In *Agroecología en el tropico: Ejemplos de Cuba*, ed. A. Leyva and J. Pohlin, 165–82. Aachen: Shaker Verlag, 2005.

Lowder, Sarah, Jacob Skoet, and Terri Raney. "The Number, Size, and Distribution of Farms, Smallholder Farms, and Family Farms Worldwide." *World Development* 87 (2016): 16–29.

Lugo, Ariel. "Comparison of Tropical Tree Plantations with Secondary Forests of a Similar Age." *Ecological Monographs* 62, no. 1 (1992): 1–41.

Lund, Jens, Rebecca Rett, and Jesse Ribot. "Trends in Research on Forestry Decentralization Policies." *Current Opinion in Environmental Sustainability* 32, no. 1 (2018): 17–22.

Lund, Jens, and Thorsten Treue. "Are We Getting There? Evidence of Decentralized Forest Management from the Tanzanian Miombo Woodlands." *World Development* 36, no. 12 (2008): 2780–800.

Maathai, Wangari. *The Green Belt Movement: Sharing the Approach and the Experience.* Herndon, VA: Lantern Books, 2004.

Maharjan, Amina, Ishaan Kochhar, Vishwas Chitale, Abid Hussain, and Giovanni Gioli. "Understanding Rural Outmigration and Agricultural Land Use Change in the Gandaki Basin, Nepal." *Applied Geography* 124 (2020): e102278.

Maher, Neil. *Nature's New Deal: The Civilian Conservation Corps and the Roots of the American Environmental Movement.* New York: Oxford University Press, 2007.

Mann, Michael. *The Sources of Social Power, Vol. 1: A History of Power from the Beginning Until 1760.* Cambridge: Cambridge University Press, 1986.

Masuda, Yuta, Brian Robinson, Margaret Holland, Tzu Tseng, and Alain Frechette. "Emerging Research Needs and Policy Priorities for Advancing Land Tenure Security and Sustainable Development." In *Land Tenure Security and Sustainable Development*, ed. Margaret Holland, Yuta Masuda, and Brian Robinson, 313–26. New York: Palgrave MacMillan, 2022.

Mather, Alexander. "Forest Transition Theory and the Reforestation of Scotland." *Scottish Geographical Journal* 120 (2004): 83–98.

——. "Recent Asian Transitions in Relation to Forest Transition Theory." *International Forestry Review* 9, no. 1 (2007): 491–502.

Mather, Alexander, and J. Fairbairn. "From Floods to Reforestation: The Forest Transition in Switzerland." *Environment and History* 6, no. 4 (2000): 399–421.

Mather, Alexander, J. Fairbairn, and C. Needle. "The Course and Drivers of the Forest Transition: The Case of France." *Journal of Rural Studies* 15, no. 1 (1999): 65–90.

Mather, Alexander, and Carolyn Needle. "The Forest Transition: A Theoretical Basis." *Area* 30, no. 2 (1998): 117–24.

Mather, Alexander, Carolyn Needle, and James Coull. "From Resource Crisis to Sustainability: The Forest Transition in Denmark." *International Journal of Sustainable Development and World Ecology* 5 (1998): 182–93.

Matricardi, Eraldo, David Skole, Olivia Costa, Marcos Pedlowski, Jay Samek, and Eder Miguel. "Long-term Forest Degradation Surpasses Deforestation in the Brazilian Amazon." *Science* 369, no. 6509 (2020): 1378–82.

Maxwell, Simon, "Marginalized Peasants to the North of Santa Cruz: Avenues of Escape from the Barbecho Crisis." In *Land, People, and Planning in Contemporary*

Amazonia, ed. F. Barbira-Scazziochio, 162–70. Cambridge: Centre for Latin American Studies, 1980.

May, Peter, Anthony Anderson, Jose Frazao, and Michael Balick. "Babassu Palm in the Agroforestry Systems in Brazil's Mid-North Region." *Agroforestry Systems* 3, no. 3 (1985): 275–95.

McDowell, N., C. Allen, K. Anderson-Teixeira, B. Aukema, B. Bond-Lamberty, L. Chini, J. Clark, et al. "Pervasive Shifts in Forest Dynamics in a Changing World." *Science* 368, no. 9463 (2020): eaaz9463.

McElvaine, Robert S. *The Great Depression: America, 1929–1941*. New York: Times Books, 1984.

McElwee, Pam. *Forests Are Gold: Trees, People, and Environmental Rule in Vietnam*. Seattle: University of Washington Press, 2016.

McGroddy, Megan, Amy Lerner, Diana Burbano, Laura Schneider, and Thomas K. Rudel. "Effects of Pasture Management on Carbon Stocks: A Study from Four Communities in Southwestern Ecuador." *Biotropica* 47, no. 4 (2015): 407–15.

McKillen, Beth. "The Corporatist Model, World War I, and the Debate About the League of Nations." *Diplomatic History* 15, no. 2 (1991): 171–97.

McNeill, John. *The Great Acceleration: An Environmental History of the Anthropocene Since 1945*. Cambridge, MA: Harvard University Press, 2016.

Meijaard, E., T. Brooks, K. Carlson, E. Slade, J. Garcia-Ulloa, D. Gaveau, J. Ser Huay Lee, T. Santika, et al. "The Environmental Impacts of Palm Oil in Context." *Nature-Plants* 6 (2020): 1418–26.

Meinizen-Dick, Ruth. "Women, Land, and Trees." In *World Agroforestry into the Future*, ed. D Garrity, A. Okono, M. Grayson, and S. Parrott, 173–79. Nairobi: World Agroforestry Centre, 2006.

Metzel, Ruth, and Florencia Montagnini. "From Farm to Forest: Factors Associated with Protecting and Planting Trees in a Panamanian Agricultural Landscape." *Bois et Forêts Tropiques* 322, no. 4 (2014): 3–15.

Meyer, John, John Boli, George Thomas, and Francisco Ramirez. "World Society and the Nation State." *American Journal of Sociology* 103, no. 1 (1997): 144–81.

Meyfroidt, Patrick, R. Roy Chowdbury, A. de Bremond, E. Ellis, T. Filatova, R. Garrett, M. Grove, A. Heinimann, et al. "Middle Range Theories of Land Use Change." *Global Environmental Change* 53 (2018): 52–67.

Meyfroidt, Patrick, and Eric Lambin. "The Causes of the Reforestation in Vietnam." *Land Use Policy* 25, no. 2 (2008): 182–97.

——. "Global Forest Transition: Prospects for an End to Deforestation." *Annual Review of Environment and Resources* 36 (2011): 343–71.

Meyfroidt, Patrick, Thomas K. Rudel, and Eric Lambin. "Forest Transitions, Trade, and the Displacement of Land Use." *Proceedings of the National Academy of Sciences* 107, no. 49 (2010): 20917–22.

Michalski, Fernanda, Jean Metzger, and Carlos Peres. "Rural Property Size Drives Patterns of Upland and Riparian Forest Retention in a Tropical Deforestation Frontier." *Global Environmental Change* 20, no. 4 (2010): 705–12.

Michon, Genevieve, Hubert de Foresta, Patrice Levang, and Francois Verdeaux. "Domestic Forests: A New Paradigm for Integrating Local Communities' Forestry into Tropical Forest Science." *Ecology and Society* 12, no. 2 (2007): 1.

Migdal, Joel. *Strong Societies and Weak States: State-Society Relationships and State Capabilities in the Developing World.* Princeton, NJ: Princeton University Press, 1988.

Mill, John Stuart. *A System of Logic: Ratiocinative and Inductive.* Toronto: University of Toronto Press, 1967.

Miteva, Daniela, Colby Loucks, and Subhrendu Pattanayak. "Social and Environmental Impacts of Forest Management Certification in Indonesia." *Plos One* 10, no. 7 (2015): e0129675.

Mol, Arthur, and David Sonnenfeld. *Ecological Modernization Around the World: Perspectives and Critical Debates.* London: Frank Cass, 2000.

Molina, Oscar, and Martin Rhodes. "Corporatism: The Past, Present, and Future of a Concept." *Annual Review of Political Science* 5 (2002): 305–31.

Molotch, Harvey. "The City as a Growth Machine: The Political Economy of Place." *American Journal of Sociology* 82, no. 2 (1976): 309–32.

Müller, Robert, Daniel Müller, Frances Schierhorn, and Gerald Gerold. "Spatiotemporal Modeling of the Expansion of Mechanized Agriculture in the Bolivian Lowland Forests." *Applied Geography* 31, no. 2 (2011): 631–40.

Muller, Daniel, Zanhli Sun, Thoumthone Vongvisouk, Dirk Pflugmacher, Jiangho Xu, and Ole Mertz. "Regime Shifts Limit the Predictability of Land-System Change." *Global Environmental Change* 28 (2014): 75–83.

Murphy, Peter, and Ariel Lugo. "Ecology of Tropical Dry Forest." *Annual Review of Ecology and Systematics* 17 (1986): 67–88.

Myers, Norman. *Conversion of Tropical Moist Forests.* Washington, DC: National Academies Press, 1980.

Nagendra, Harini, Jane Southworth, and Catherine Tucker. "Accessibility as a Determinant of Landscape Transformation in Western Honduras: Linking Pattern and Process." *Landscape Ecology* 18 (2003): 141–58.

Nanni, A. Sofia, and H. Ricardo Grau. "Agricultural Adjustment, Population Dynamics and Forest Redistribution in a Subtropical Watershed of NW Argentina." *Regional Environmental Change* 14 (2014): 1641–49.

National Academies of Sciences, Engineering, and Medicine. *Negative Emissions Technologies and Reliable Sequestration: A Research Agenda.* Washington, DC: National Academies Press. 2019.

Nations, James, and Daniel Komer. "Rain Forests and the Hamburger Society." *Environment* 25, no. 3 (1983): 12–20.

Naylor, Rosamond, Mathew Higgins, Ryan Edwards, and Walter Falcon. "Decentralization and the Environment: Assessing Smallholder Oil Palm Development in Indonesia." *Ambio* 48 (2019): 1195–1208.

Netting, R. M. *Smallholders, Householders: Farm Families and the Ecology of Intensive, Sustainable Agriculture.* Stanford, CA: Stanford University Press, 1993.

Niether, Wiebke, Johanna Jacobi, Wilma Blaser, Christian Andres, and Laura Armengot. "Cocoa Agroforestry Systems Versus Monocultures: A Multi-Dimensional Meta-Analysis." *Environmental Research Letters* 15 (2020): 104085.

NYDF Assessment Partners. "Protecting and Restoring Forests: A Story of Large Commitments Yet Limited Progress—New York Declaration on Forests Five-Year Assessment Report." 2019. www.forestdeclaration.org.

Oberg, Perola, Torsten Svensson, Peter Christiansen, Asbjorn Norgaard, Hilmar Rommetvedt, and Gunnar Thesen. "Disrupted Exchange and Declining Corporatism." *Government and Opposition* 46, no. 3 (2011): 365–91.

Odum, Howard. *The Southern Regions of the United States*. Chapel Hill: University of North Carolina Press, 1936.

Oldekop, Johan, Katherine Sims, Mark Whittingham, and Arun Agrawal. "International Migration Drives Reforestation in Nepal." *Global Environmental Change* 52 (2018): 66–74.

Oliveira, Tiago, Nun Guiomar, Fernando Baptista, Jose Pereira, and Joao Claro. "Is Portugal's Forest Transition Going up in Smoke?" *Land Use Policy* 66 (2017): 214–26.

Omont, Hubert, Dominique Nicolas, and Diane Russell. 2006. "The Future of Perennial Tree Crops: What Role for Agroforestry?" In *World Agroforestry into the Future*, ed. D. Garrity, A. Okono, M. Grayson, and S. Parrott, 23–35. Nairobi: World Agroforestry Centre, 2006.

O'Neill, Karen. *Rivers by Design: State Power and the Origins of U.S. Flood Control*. Durham, NC: Duke University Press, 2006.

Ordway, Elsa, Greg Asner, and Eric Lambin. "Deforestation Risk Due to Commodity Crop Expansion in Sub-Saharan Africa." *Environmental Research Letters* 12 (2017): 044015.

Ortolani, Giovanni "Agroforestry: An Increasingly Popular Solution for a Hot, Hungry World." *Mongabay*, October 26, 2017.

Ostrom, Elinor. "Beyond Markets and States: Polycentric Governance of Complex Economic Systems." *American Economic Review* 100 (2010): 641–72.

——. *Governing the Commons: The Evolution of Institutions for Collective Action*. New York: Cambridge University Press, 1990.

——. "Polycentric Systems for Coping with Collective Action and Global Environmental Change." *Global Environmental Change* 20 (2010): 550–57.

Oswalt, Sonja, W. Brad Smith, Patrick Miles, and Scott Pugh. *Forest Resources of the United States*. Washington, DC: United States Department of Agriculture, 2017.

Otero, Luis. *La huella del fuego: Historia de los bosques nativos—Poblamiento y cambios en el paisaje del sur de Chile*. Santiago, Chile: Pehuén Editores Limitada, 2006.

Pearce, Fred. "Sparing Versus Sharing: The Great Debate Over How to Protect Nature." *Environment 360*.2018. https://e360.yale.edu/features/sparing-vs-sharing-the-great-debate-over-how-to-protect-nature.

——. *A Trillion Trees: How We Can Reforest Our World*. London: Granta, 2021.

Peluso, Nancy. "Coercing Conservation: The Politics of State Resource Control." *Global Environmental Change* 3, no. 2 (1993): 199–217.

Phalan, Ben, Malvika Onial, Andrew Balmford, and Rhys Green. "Reconciling Food Production and Biodiversity Conservation: Land Sharing and Land Sparing Compared." *Science* 333 (2011): 1289–91.

Philips, Sarah. *This Land, This Nation: Conservation, Rural America, and the New Deal*. New York: Cambridge University Press, 2007.

Piketty, Thomas, and Emmanuel Saez. "Inequality in the Long Run." *Science* 344, no. 6186 (2014): 838–43.

Pinedo-Vasquez, Miguel, Daniel Zarin, Kevin Coffey, Christine Padoch, and Fernando Rabelo. "Post-boom Timber Production in Amazonia." *Human Ecology* 29 (2001): 219–39.

Poffenberger, Mark, and Betsy McGean. *Village Voices, Forest Choices: Joint Forest Management in India.* Delhi: Oxford University Press, 1996.

Polain de Waroux, Yann, M. Baumann, N. Ignacio Gasparri, G. Gavier-Pizarro, J. Godar, T. Kuemmerle, R. Müller, et al. "Rents, Actors, and the Expansion of Commodity Frontiers in the Gran Chaco." *Annals of the American Association of Geographers* 108, no. 1 (2018): 204–25.

Polain de Waroux, Yann, Rachael Garrett, Jordan Graesser, Christoph Nolte, Christopher White, and Eric Lambin. "The Restructuring of South American Soy and Beef Production and Trade Under Changing Environmental Regulations." *World Development* 121 (2019): 188–202.

Polain de Waroux, Yann, Rachael Garrett, Robert Heilmayr, and Eric Lambin. "Land-use Policies and Corporate Investments in Agriculture in the Gran Chaco and Chiquitano." *Proceedings of the National Academy of Sciences* 113, no. 15 (2016): 4021–26.

Polanyi, Karl. *The Great Transformation.* New York: Farrar and Rinehart, 1944.

Porro, Roberto. "Land Use, Cattle Ranching, and the Concentration of Landownership in Maranhao, Brazil." In *Deforestation and Land Use in the Amazon*, ed. Charles H. Wood and Roberto Porro, 315–37. Gainesville: University of Florida Press, 2002.

——. "Palms, Pastures, and Swidden Fields: The Grounded Political Ecology of Agro-Extractive/Shifting-Cultivator Peasants in Maranhao, Brazil." *Human Ecology* 33, no. 1 (2005): 17–56.

Post, Eric, and Michelle Mack. "Arctic Wildfires at a Warming Threshold. *Science* 378, no. 6619 (2022): 470.

Potapov, P., M. Hansen, L. Laestaduis, S. Turubanova, A. Yaroshenko, C. Thies, W. Smith, et al. "The Last Frontiers of Wilderness: Tracking Loss of Intact Forest Landscapes from 2000 to 2013." *Science Advances* 3, no. 1 (2017): 1600821.

Potts, Jason, Matthew Lynch, Ann Wilkings, Gabriel Huppé, Maxine Cunningham, and Vivek Vorra. "The State of Sustainability Initiatives Review 2014: Standards and the Green Economy." International Institute for Sustainable Development (IISD) and the International Institute for Environment and Development (IIED). 2014. https://www.iisd.org/publications/state-sustainability-initiatives-review -2014-standards-and-green-economy.

Preston, David. "Changed Household Livelihood Strategies in the Cordillera of Luzon." *Tijdschrift voor Economische en Sociale Geografie* 89, no. 4 (1998): 371–83.

Pulhin, Florencia, Rodel Lasco, and Joan Urquiola. "The Carbon Sequestration Potential of Oil Palm in Bohol, Philippines." *Ecosystems and Development Journal* 4, no. 2 (2014): 14–19.

Pye-Smith, Charlie. "Rich Rewards for Rubber? Research in Indonesia Is Exploring How Smallholders Can Increase Rubber Production, Retain Biodiversity and Provide Additional Environmental Benefits." Nairobi: World Agroforestry Centre, 2011.

Ragin, Charles. *The Comparative Method: Moving Beyond Qualitative and Quantitative Strategies.* Berkeley: University of California Press, 1987.

Rajão R., B. Soares-Filho, F. Nunes, J. Börner, L. Machado, D. Assis, A. Oliveira, et al. "The Rotten Apples of Brazil's Agribusiness: Brazil's Inability to Tackle Illegal Deforestation Puts the Future of Its Agribusiness at Risk." *Science* 369, no. 6501 (2020): 246–48.

Ramankutty, Navin, and Oliver Coomes. "Land-use Regime Shifts: An Analytical Framework and Agenda for Future Land Use Research." *Ecology and Society* 21, no. 2 (2016): 1.

Ramankutty, Navin, Elizabeth Heller, and Jeanine Rhemtulla. "Prevailing Myths About Agricultural Abandonment and Forest Regrowth in the United States." *Annals of the Association of American Geographers* 100, no. 3 (2010): 502–12.

Raper, Arthur. *A Preface to Peasantry: A Tale of Two Black Belt Counties.* Chapel Hill: University of North Carolina Press, 1936.

——. *Tenants of the Almighty.* New York: Macmillan, 1943.

Raper, Arthur, and Ira Reid. *Sharecroppers All.* New York: Russell & Russell, 1941.

Raup, Hugh. "The View from John Sanderson's Farm: A Perspective for the Use of Land." *Forest History* 10, no. 1 (1966): 2–11.

Redo, Daniel, H. Ricardo Grau, T. Mitchell Aide, and Matthew Clark. "Asymmetric Forest Transition Related to the Interaction of Socio-economic Development and Forest Type in Central America." *Proceedings of the National Academy of Sciences* 109, no. 23 (2012): 8839–44.

Reid, J. Leighton, Matthew Fagan, James Lucas, and Joshua Slaughter. "The Ephemerality of Secondary Forests in Southern Costa Rica." *Conservation Letters* 12 (2019): e12607.

Reid, John W., and Thomas Lovejoy. *Ever Green: Saving Big Forests to Save the Planet.* New York: Norton, 2022.

Reij, Chris, Gray Tappan, and Melinda Smale. "Agroenvironmental Transformation in the Sahel: Another Kind of 'Green Revolution.' " Washington, DC: International Food Policy Research Institute, 2009.

Ritchie, Hannah, and Max Roser. "Protected Areas and Conservation." Our World in Data. 2021. https://ourworldindata.org/biodiversity.

Rival, Alain, and Patrice Levang. *Palms of Controversies: Oil Palm and Development Challenges.* Bogor, Indonesia: CIFOR, 2014.

Rochmyaningsih, Dyna. "Courting Controversy, Scientists Team with Industry to Tackle One of the World's Most Destructive Crops." *Science* 365, no. 6649 (2019): 112–15.

——. "Massive Road Project Threatens New Guinea's Biodiversity." *Science* 374, no. 6565 (2021): 246–47.

Rudel, Thomas K. *Defensive Environmentalists and the Dynamics of Global Reform.* New York: Cambridge University Press, 2013.

——. "Did a Green Revolution Reforest the American South?" In *Agricultural Technologies and Tropical Deforestation,* ed. A. Angelsen and D. Kaimowitz, 33–54. New York: CABI Press, 2001.

——. "The Dynamics of Deforestation in the Wet and Dry Tropics: A Comparison with Policy Implications." *Forests* 8, no. 4 (2017): 108.

——. "The Extractive Imperative in Populous Indigenous Territories: The Shuar, Copper Mining, and Environmental Injustices in the Ecuadorian Amazon." *Human Ecology* 46, no. 5 (2018): 717–24.

——. "Indigenous-Driven Sustainability Initiatives in Mountainous Regions: The Shuar in the Tropical Andes of Ecuador." *Mountain Research and Development* 41, no. 1 (2021): 22–28.

——. "Shocks, States, and Societal Corporatism: A Shorter Path to Sustainability?" *Journal of Environmental Studies and Sciences* 9, no. 4 (2019): 429–36.

——. *Shocks, States, and Sustainability: The Origins of Radical Environmental Reforms.* New York: Oxford University Press, 2019.

——. "The Variable Paths to Sustainable Intensification in Agriculture." *Regional Environmental Change* 20, no. 126 (2020): 126.

Rudel, Thomas, Greg Asner, Ruth Defries, and William Laurance. "Changing Drivers of Deforestation and New Opportunities for Conservation." *Conservation Biology* 23 (2009): 1396–1405.

Rudel, Thomas, and Bruce Horowitz. *Tropical Deforestation: Small Farmers and Land Clearing in the Ecuadorian Amazon.* New York: Columbia University Press, 1993.

Rudel, Thomas K., and Monica Hernandez. "Land Tenure Transitions in the Global South: Trends, Drivers, and Policy Implications." *Annual Review of Environment and Resources* 42 (2017): 489–507.

Rudel, Thomas K., Patrick Meyfroidt, Ruth Chazdon, Frans Bongers, Sean Sloan, H. Ricardo Grau, T. Van Holt, and Laura Schneider. "Whither the Forest Transition? Climate Change, Policy Responses, and Redistributed Forests in the Twenty-First Century." *Ambio* 49, no. 1 (2020): 74–84.

Rudel, Thomas, Karen O'Neill, Paul Gottlieb, Melanie McDermott, and Colleen Hatfield. "From Middle to Upper Class Sprawl? Land Use Controls and Real Estate Development in Northern New Jersey." *Annals of the Association of American Geographers* 101 (2011): 609–24.

Rueda, Ximena and Eric Lambin. "Responding to Globalization: Impacts of Certification on Colombian Small-scale Coffee Growers." *Ecology and Society* 18, no. 3 (2013): 21.

Ruf, François. "The Myth of Complex Cocoa Agro-Forests." *Human Ecology* 39 (2011): 373–88.

——. "Tree Crops as Deforestation and Reforestation Agents: The Case of Cocoa in Côte d'Ivoire and Sulawesi." In *Agricultural Technologies and Tropical Deforestation*, ed. A. Angelsen and D. Kaimowitz, 291–315. Wallingford, UK: CABI Publishing, 2001.

Saloutos, Theodore. *The American Farmer and the New Deal.* Ames: Iowa State University Press, 1982.

Samberg, Leah, James Gerber, Navin Ramankutty, Mario Herrero, and Paul West. "Subnational Distribution of Average Farm Size and Smallholder Contributions to Global Food Production." *Environmental Research Letters* 11, no. 12 (2016): 124010.

Schelhas, John. "Land Use Choice and Change: Intensification and Diversification in the Lowland Tropics of Costa Rica." *Human Organization* 55, no. 3 (1996): 298–306.

Schelhas, John, Tom Brandeis, and Thomas K. Rudel. "Reforestation and Natural Regeneration in Forest Transitions: Patterns and Implications from the U.S. South." *Regional Environmental Change* 21, no. 1 (2021): 8.

Schelhas, John, and Max Pfeffer. *Saving Forests, Protecting People: Environmental Conservation in Central America*. Lanham, MD: Altamira Press, 2008.

Schmitt-Harsh, Mikaela, Tom Evans, Edwin Castellanos, and James Randolph. "Carbon Stocks in Coffee Agroforests and Mixed Dry Tropical Forests in the Western Highlands of Guatemala." *Agroforestry Systems* 86 (2012): 141–57.

Schmitter, Philippe. "Still a Century of Corporatism?" *Review of Politics* 36, no. 1 (1974): 85–131.

Schroth, Gotz, Deborah Faria, Marcelo Araujo, Lucio Bede, Sunshine Van Bael, Camilla Cassano, Leonardo Oliveira, and Jacques Delabie. "Conservation in Tropical Landscape Mosaics: The Case of the Cacao Landscape of Southern Bahia, Brazil." *Biodiversity and Conservation* 20 (2011): 1635–54.

Schwartz, Naomi, T. Mitchell Aide, Jordan Graesser, H. Ricardo Grau, and Maria Uriarte. "Reversals of Reforestation Across Latin America Limit Climate Mitigation Potential of Tropical Forests." *Frontiers in Forests and Global Change* 3 (2020): 85.

Scott, James. *Seeing Like a State: How Certain Schemes to Improve the Human Condition Have Failed*. New Haven, CT: Yale University Press, 1998.

Sen, Amarta. "The Ends and Means of Sustainability." *Journal of Human Development and Capabilities* 14, no. 1 (2013): 6–20.

Sendzimir, Jan, Chris Reij, and Piotr Magnuszewski. "Rebuilding Resilience in the Sahel: Regreening in the Maradi and Zinder Regions of Niger." *Ecology and Society* 16, no. 3 (2011): 1.

Seymour, Frances, and Jonas Busch. *Why Forests? Why Now? The Science, Economics, and Politics of Tropical Forests and Climate Change*. Washington, DC: Center for Global Development, 2016.

Shane, Douglas R. *Hoofprints in the Forest: Cattle Ranching and the Destruction of the Amazon Forest*. Philadelphia: Institute for the Study of Human Issues, 1986.

Sikor, Tomas, G. Auld, A. Bebbington, T. Benjaminsen, B. Gentry, C. Hunsberger, A-M. Izac, M. Margulis, et al. "Global Land Governance: From Territory to Flow." *Current Opinion in Environmental Sustainability* 5, no. 5 (2013): 522–27.

Sikor, Tomas, and Jacopo Baggio. "Can Smallholders Engage in Tree Plantations? An Entitlements Analysis from Vietnam." *World Development* 64 (2014): 101–12.

Sikor, Tomas, Nghiem Tuyen, Jennifer Sowerwine, and Jeff Romm. *Upland Transformations in Vietnam*. Singapore: National University of Singapore Press, 2011.

Sloan, Sean, Patrick Meyfroidt, Thomas K. Rudel, Frans Bongers, and Ruth Chazdon. "The Forest Transformation: Planted Tree Cover and Regional Dynamics of Tree Gains and Losses." *Global Environmental Change* 59 (2019): 101988.

Song, X., et al. "Global Land Change from 1982 to 2016." *Nature* 560 (2018): 639–44.

Sonwa, D., B. Nkongmeneck, S. Weise, M. Tchatat, A. Adesina, M. Janssens, et al. "Diversity of Plants in Cocoa Agroforests in the Humid Forest Zone of Southern Cameroon." *Biodiversity and Conservation* 16, no. 8 (2007): 2385–400.

Strassburg, Bernardo, et al. "Global Priority Areas for Ecosystem Restoration." *Nature* 586 (2020): 724–29.

Streeck, Wolfgang, and Lane Kenworthy. "Theories and Practices of Neo-corporatism." In *The Handbook of Political Sociology: States, Civil Societies, and Globalization*, ed. T. Janoski, R. Alford, A. Hicks, and M. Schwartz, 441–60. Cambridge: Cambridge University Press, 2005.

Stuart, Diana, Ryan Gunderson, and Brian Petersen. *The Degrowth Alternative: A Path to Address Our Environmental Crisis?* New York: Routledge, 2021.

Sunderlin, W., C. de Sassi, E. Sills, A. Duchelle, A. Larson, et al. "Creating an Appropriate Tenure Foundation for REDD+: The Record to Date and Prospects for the Future." *World Development* 106 (2018): 376–92.

Sunderlin, W., A. Ekuputri, E. Sills, A. Duchelle, D. Kweka, R. DiProse, N. Doggart, et al. "The Challenge of Establishing REDD+ on the Ground: Insights from 23 Subnational Initiatives in Six Countries." Bogor, Indonesia: Center for International Forestry Research, 2014.

Tacconi, L. "Fires in Indonesia: Causes, Costs, and Policy Implications." Bogor, Indonesia: Center for International Forestry Research, 2003.

Terborgh, John. *Requiem for Nature*. Washington, DC: Island Press, 1999.

Thunberg, Greta. *No One Is Too Small to Make a Difference*. New York: Penguin Books, 2019.

Tiffen, Mary, Michael Mortimore, and Frances Gichuki. *More People, Less Erosion: Environmental Recovery in Kenya*. Chichester, UK: Wiley, 1993.

Tompkins, Lucy. "What Does It Mean to Sell Forest Candy?" *New York Times*, December 5, 2021.

Truong, Dao, Masayuki Yanagisawa, and Yasuyuki Kono. "Forest Transition in Vietnam: A Case Study of Northern Mountain Region." *Forest Policy and Economics* 76 (2017): 72–80.

Tseng, T., B. Robinson, M. Bellemare, A. BenYishay, A. Blackman, T. Boucher, M. Childress, et al. "Influence of Land Tenure Interventions on Human Well-Being and Environmental Outcomes." *Nature Sustainability* 4 (2021): 242–51.

United Nations, Climate Secretariat. *Intended Nationally Determined Contributions*. New York: United Nations, 2016.

Vale, Petterson, Holly Gibbs, Ricardo Vale, Matthew Christie, Eduardo Florence, Jacob Munger, and Derquaine Sabaini. "The Expansion of Intensive Beef Farming to the Brazilian Amazon." *Global Environmental Change* 57 (2019): 101922.

Vancutsem, C., F. Achard, J-F. Pekel, G. Vieilledent, S. Carboni, D. Simonetti, J. Gallego, et al. "Long-term (1990–2019) Monitoring of Forest Cover Changes in the Humid Tropics." *Science Advances* 7 (2021): eabe1603.

Van Noordwijk, Meine, Ric Coe, and Fergus Sinclair. "Agroforestry Paradigms." In *Sustainable Development Through Trees on Farms: Agroforestry in Its Fifth Decade*, ed. M. van Noordwijk, 1–14. Bogor, Indonesia: World Agroforestry (ICRAF) Southeast Asia Regional Program, 2019.

Vayda, Andrew P., and Ahmad Sahur. "Forest Clearing and Pepper Farming by Bugis Migrants in East Kalimantan: Antecedents and Impacts." *Indonesia* 39, no. 4 (1985): 93–110.

Von Thünen, Johan. *Von Thünen's Isolated State*. Trans. Carla M. Wartenberg. London: Pergamon Press, 1967.

Vyawahare, Malavika. "Conflict and Climate Change Are Big Barriers for Africa's Great Green Wall." *Mongabay*, November 26, 2021.

Wakker, Eric. *The Kalimantan Border Oil Palm Mega-Project*. Amsterdam: Friends of the Earth—Netherlands, 2006.

Walker, Nathalie, Sabrina Patel, and Kemel Kalif. "From Amazon Pasture to the High Street: Deforestation and the Brazilian Cattle Product Supply Chain." *Tropical Conservation Science* 6, no. 3 (2013): 446–67.

Walters, Bradley. *The Greening of Saint Lucia: Economic Development and Environmental Change in the Eastern Caribbean*. Kingston, Jamaica: University of the West Indies Press, 2019.

——. "Migration, Land Use, and Forest Change in St. Lucia, West Indies." *Land Use Policy* 51 (2016): 290–300.

Wessel, Marius, and P. Oquist-Wessel. "Cocoa Production in West Africa, a Review and Analysis of Recent Developments." *NJAS—Wageningen Journal of the Life Sciences* 74–75 (2015): 1–7.

Wibawa, Gede, Sinung Hendratno, and Meine van Noordwijk. "Permanent Smallholder Rubber Agroforestry Systems in Sumatra, Indonesia." In *Slash and Burn Agriculture: The Search for Alternatives*, ed. C. Palm, S. Vosti, P. Sanchez, and P. Ericksen, 222–32. New York: Columbia University Press, 2005.

Widener, Patricia. *Oil Injustice: Resisting and Conceding a Pipeline in Ecuador*. Lanham, MD: Rowman & Littlefield, 2011.

Wilkerson, Isabel. *The Warmth of Other Suns: The Epic Story of America's Great Migration*. New York: Random House, 2010.

Williams, Michael. *Deforesting the Earth: From Prehistory to Global Crisis*. Chicago: University of Chicago Press, 2006.

Wilson, Edward O. *Half-Earth: Our Planet's Fight for Life*. New York: Norton, 2016.

Wilson, Sarah, John Schelhas, H. Ricardo Grau, Sofia Nanni, and Sean Sloan. "Forest Ecosystem Transitions: The Ecological Dimensions of the Forest Transition." *Ecology and Society* 22, no. 4 (2017): 38.

Winkler, Karina, Richard Fuchs, Mark Rounsevell, and Martin Herold. "Global Land Use Changes Are Four Times Greater Than Previously Estimated." *Nature Communications* 12 (2021): 2501.

Wolf, Christopher, Taal Levi, William Ripple, Diego Zárrate-Charry, and Matthew Betts. "A Forest Loss Report Card for the World's Protected Areas." *Nature Ecology and Evolution* 5, no. 4 (2021): 520–29.

Wolosin, Michael. *Large-Scale Forestation for Climate Mitigation: Lessons from South Korea, China, and India*. Washington, DC: Climate and Land Use Alliance, 2017.

Woodruffe, B. J. "Rural Land Use Planning in West Germany." In *Rural Land Use Planning in Developed Nations*, ed. Paul J. Cloke, 104–29. London: Unwin Hyman, 1989.

World Bank. *State and Trends of Carbon Pricing, 2020*. Washington, DC: World Bank, 2020.

World Resources Institute (WRI). CAIT Climate Data Explorer. 2018. http://cait.wri.org/indc/#/.

Wrigley, Edward. *Population and History*. London: Weidenfeld and Nicolson, 1969.

Wunder, Sven, Amy Duchelle, Claudio de Sassi, Erin Sills, Gabriela Simonet, and William Sunderlin. "REDD+ in Theory and Practice: How Local Projects Can Inform Jurisdictional Approaches." *Frontiers in Forests and Global Change* 3, no. 11 (2020). https://www.cifor.org/publications/pdf_files/articles/AWunder2001.pdf.

York, Richard. "Ecological Paradoxes: William Stanley Jevons and the Paperless Office." *Human Ecology Review* 13 (2006): 141–46.

Ziegler, Alan, J. Phelps, J. Yuen, E. Webb, D. Lawrence, J. Fox, T. Bruun, et al. "Carbon Outcomes of Major Land-Cover Transitions in SE Asia: Great Uncertainties and REDD+ Policy Implications." *Global Change Biology* 18 (2012): 3087–99.

Ziegler, Alan, Jefferson Fox, and Jiancho Xu. "The Rubber Juggernaut." *Science* 324 (2009): 1024–25.

Zinda, John, Christine Trac, Deli Zhai, and Stevan Harrell. "Dual Function Forests in the Returning Forests to Farmland Program and Flexibility in Environmental Policy in China." *Geoforum* 78 (2017): 119–32.

Zomer, Robert, Henry Neufeldt, Jiancho Xu, Antje Ahrends, Deborah Bossio, Antonio Trabucco, Meine van Noordwijk, and Mingcheng Wang. "Global Tree Cover and Biomass Carbon on Agricultural Land: The Contribution of Agroforestry to Global and National Carbon Budgets." *Science Reports* 6 (2016): 29987.

Zomer, Robert, Antonio Trabucco, Ric Coe, and Frank Place. "Trees on Farms: Analysis of Global Extent and Geographical Patterns of Agroforestry." Nairobi: World Agroforestry Centre, 2009.

Zomer, Robert, Antonio Trabucco, Ric Coe, Frank Place, Meine van Noordwijk, and Jiancho Xu. "Trees on Farms: An Update and Reanalysis of Agroforestry's Global Extent and Socio-Ecological Characteristics." Bogor, Indonesia: World Agroforestry Centre (ICRAF) Southeast Asia Regional Program, 2014.

Index

Africa, sub-Saharan: and deforestation around cities, 144; and dry forests, 141–46

African Americans, 77–80, 116

agricultural expansion, 13–17, 35

agricultural intensification, 165; in Costa Rica, 84; in forest plantations, 124–25; in the southern United States, 81–82

agricultural lands, as sites for forest plantations in Mozambique, 122

agroforests, 149–74; cocoa, 158–64; coffee, 164–68; diversified production, 149–52; extent of cultivation, 150; jungle rubber, 151–57; numbers of cultivators, 150

Aichi Targets, 5

Argentina, 63

avoided deforestation, 42–67

Babassu palms, 137–41; economic uses of, 138–39

Barbecho crisis, in Bolivia132

Bolsonaro, Jair, 64

Bonn Challenge, 5, 27, 100

boreal forests, 3, 188–89

Boserup, Ester, 36

Brazil, 30, 53–65, 137–41, 160; and Amazon fund, 59; intensified forest plantations in, 124–25. *See* Bolsonaro, Jair; Cardoso, Fernando Henrique; Lula da Silva, Luiz Inácio

Caboclos, and Acai regrowth and trade, 99, 178

cacao, 158–64; cultivation of Cote d'Ivoire, 159; extension service efforts, 161–62; grown in full sun, 158; migrant cultivators of in Cote d'Ivoire, 159; shade-grown, 158–60; trends in area cultivated, 158. *See also* Sustainable Cocoa Production Programme (SCPP)

Cameroon, 31, 160

Campaign for Nature, 6

carbon offsets, 29–31, 136

carbon sequestration, xi, 1–3, 30–31, 121, 127; in silvopastures, 141; on tree farms as carbon sinks, 97, 121; and tropical rain forests, 3

Cardoso, Fernando Henrique, 62

case studies, xii, 10–11. 40–41

cattle ranching, 131–41; and declines in price of beef, 83–84; and pasture grasses, 132; and ranchers in Brazil, 137–41

Central America, 12

Chile, 116–19; expansions in plantation forests in, 117–18; historical fluctuations in forest cover in, 117

China: crisis narratives in, 103; Returning Farmland to Forest Program (RFFP, Grain for Green Program), 102–6. *See also* state corporatism

class conflict, in silvopastures, 137–41, 185

climate change, xi, 1–2; and fires, 2

Climate Change Framework Convention: Glasgow and Paris Conferences of Parties, 211–12; nationally determined contributions (NDCs) from the Paris COP, 119

coffee agroforests, 164–67; conversion from primary and secondary rain forest to, 164. *See* ecocertification

collective action, 7, 26, 139. *See* corporatism; social movements; countermovements

colonial expansion, 14–17

community forest enterprises, 8, 19, 55–57

conjunctural causation, 40–41

conservation movement, 33–35, 42–67; birders in Mindo, Ecuador, 52

Convention on Biological Diversity (1992), 5, 22; Aichi Targets established at, 5

Cordillera de Kutukú, 45–48

Cordillera del Cóndor, 45–48

corporatism, 9–10, 22–29, 179–86, 205–11; and concertation, 24, 27; and negotiations between different strata of society, 24–27; and norms about how to do things, 26; and sectoral associations, 24. *See* societal corporatism; state corporatism

Costa Rica, 53, 83–85

Cote d'Ivoire, 159–64; cacao agroforestry, 158–60

cotton belt, in American South, 77–82

countermovements, 20–22, 178–179

crisis narratives, 121; in China, 103. *See* floods

Cuba, 95

Davis, Kingsley, 212

deforestation, xii, 8, 13, 47–49; trends in, 13

degrowth, 198, 214

Democratic Republic of the Congo (DRC), 30, 101–2

Depression of the 1930s, 21, 80

domestic (or kitchen) forests, 168–71; and smallholder households, 168–71; gender and tree planting, 169; NGOs and tree planting campaigns, 170

double movement, 20–22, 28, 197. *See* Polanyi, Karl

dry forests, in sub-Saharan Africa, 141–46

Durkheim, Émile, 22

Earth Summit (Rio de Janeiro, 1992), 5, 22

ecocertification, 31–32, 165–74, 210–11; certified versus non-certified growers, 113, 165; meta-analysis of effects of, 113, 167; and price differentials, 167; waves of, 166

Ecological Focus Area, European Union, 93

Ecosystem Restoration Decade, United Nations, 5

ecotourism, 53, 85

Ecuador, 42–48, 131–37

eucalyptus, 4, 117

farmers, 6, 8, 13, 27, 35–40; in the Cotton Belt in the United States, 77–82; in Europe, 73–76; sharecroppers, 77; shifting cultivators in East Asia, 151–52; tenant, 78 fires in forests, 2–3, 125–26; in the American West, 3; in boreal forests, 3; in Portugal, 76; in Southeast Asia, 2

floods, 94, 102–3

Food and Agricultural Organization of the United Nations (FAO), 5, 146

food security, 122

forests, effects of climate on, 2; forest restoration, 68–96. See also dry forests, in sub-Saharan Africa; managed forests; planted forests;

forest area, 12,13; global trends in, 2, 13; in Portugal, 73

forest concessions: in the Congo, 100–101; in Indonesia, 98; in Laos, 109–10

forest cover trends, 2, 13–14; differences in between the Global North and the Global South, 200

forest plantations; 97–127; as carbon sinks vs. carbon sources, 105, 121; in Chile, 118–20; expansion of as a driver of old growth deforestation, 113, 117; growth in the American South, 114–15; harvesting rates, 100, 121, 125; locations of, 99–100; and plantations as sources for biofuels, 115; and plantation understories, 124–25; rate of timber product removals, 105; small-scale, 102–12, 123

forest restoration, 35–40, 68–96; length of time for regrowth and, 85–86

forest transitions, xii, 7, 193–99; differences between early and late transitions, 200; latecomer transitions, 200. See global forest transitions; just forest transitions

Forest Stewardship Council (FSC), 101, 118; in Chile,119; in the DRC, 101; and Greenpeace, 102

France, immigration to, 74

frontier forests, 13

fuelwood: around cities, 49, 144; in woodlands and parklands of Sub-Saharan Africa, 141–46

Gerschenkron, Alexander, 199–200

Global Environmental Facility, 9, 136

global forest transitions, 198–216

Gordon and Betty Moore Foundation, 34

Great Depression. See Depression of the 1930s

great economic acceleration, 14–15

Great Green Wall Initiative, 146

Great Transformation, 14, 20. See Polanyi, Karl

greenhouse gas emissions, 4, 14, 27, 49, 60, 66, 105, 140; across the globe, 201; in the American South, 115; in Chile, 119; in Indonesia, 110–12

Greenpeace, 62, 102

growth coalitions, 16–17, 177; and deforestation, 16–17

illegal logging, 111–12

Indigenous peoples, 17, 23, 34, 53–55; in Ecuador, 42–48; global movement of, 55, 206–7; and titles to land, 45

Indonesia, 30, 64, 110–14; illegal logging in, 112; oil palm cultivation in, 111–14; REDD+ agreement with Norway, 60; state strength in, 110

inequality, economic, 15

Initiative 20x20, 7, 9, 29

Intergovernmental Panel on Climate Change (IPCC), 5, 22, 187; conferences of parties in Glasgow and Paris, 5–6, 120, 187

Intergovernmental Science-Policy Platform on Biodiversity and Ecosytem Services (IPBES), 202

intermediary groups, 21–22. See Durkheim, Émile

international trade, 15–16, 104; and foodstuffs to Europe, 89

International Union for the Conservation of Nature, 21

just forest transitions, 123, 213–16

kitchen or domestic forests, 168–71;
   and smallholder households,
   168–71; gender and tree planting,
   169; NGOs and tree planting
   campaigns, 170

labor market dynamics, 74–75, 87–88,
   178
land-change scientists, xi, 126; remote
   sensing analyses, 18
land grabs: in the DRC, 101–2; in
   Indonesia, 110; in Laos, 109
landholders, links between large and
   small, 112, 115
landowners: large farms and land
   abandonment, 89–91; in large and
   small farm districts, 17–18; large
   and small landowners in Portugal,
   73–76; small landowner districts,
   17–18. See domestic (kitchen) forests
landscapes sector, 27; and corporatism,
   27
land sparing, 68–96
land tenure, 30–31, 90, 143–44, 169,
   171; and insecurity in Babassu zones
   of Brazil, 137–41; and REDD+, 30–31
Laos, forest concessions in, 109–10
leaf area, growth, 105
leakage, of land use controls, 63
logging, 98; and illegal logging, 111
Lula da Silva, Luiz Inácio, 59

managed forests, 201
Melanesia, 14
Mexico, regional reforestation patterns
   in, 8, 55–57. See community forest
   enterprises
Miombo woodlands, 142; reforestation
   initiative in, 144
Moore Foundation. See Gordon and
   Betty Moore Foundation
mountains, and forestation, 13,
   189–90; terrain and reforestation in
   southern Illinois, 92

Nationally Determined Contributions
   (NDCs), 119, 187, 200–201, 204,
   212–213
natural climate solutions, 1, 3–4,7–8,
   11. See also carbon sequestration
New Deal, domestic allotment program,
   81; See Depression of the 1930s
New York Declaration on Forests, 5, 27
Niger, 141–46
nongovernmental organizations
   (NGOs), 7, 20–27; and
   ecocertification, 113; and timber
   companies in Chile, 118–19
Norway, 29, 57–61

oil palm: and ecocertification, 113–14;
   in Indonesia, 91; land use
   conversion with rubber, 156
Ostrom, Elinor, 196, 212

Padua, Maria, 50
parklands and woodlands, 141–46;
   drought in, 141–42; and population
   growth, 143. See Miombo
   woodlands
pastures, in the Amazon, 128–48;
   cleaning the pastures of seedlings,
   135
payments for environmental services
   (PES), 57–61, 123. See also reduced
   emissions from deforestation and
   degradation (REDD+)
planted forests, 97–127
Polanyi, Karl, 14–15, 20. See also double
   movement; Great Transformation
polycentrism, 195–97, 212–13
population growth, 14, 198
Portugal, 73–76
poverty, rural, 130; and dry forests,
   141–42; fuelwood as alleviating, 130
precipitation, 2, 12; and droughts, 2,
   141–42; and forest fires, 2; and
   forest growth, 2, 13; trends in, 2
protected areas, 33–35, 42–67, 84

qualitative comparative analysis, 180–86

reduced emissions from deforestation and degradation (REDD+), 21, 29–31, 123, 208–10; and corporatism, 29–31; and trends towards jurisdictional programs, 208. *See* payments for environmental services (PES)

regional conservation coalitions, 51–53, 206; in Ecuador, 51; in northeastern United States, 52

regional reforestation patterns, 19, 186–95; Amazon silvopastures, 192–93; community forests, 195; dense agroforests, 193–94; dry domestic forests, 194–95; intact forests, 188–189; large forest plantations, 191–92; montane reforestation, 189–90; smallholder forest plantations, 190–91

regions of refuge, 17

remittances: and boomerang return migration, 74–76; from labor migrants, 74–76; and retirement homes, 74–76

resource-partitioning, 51; in the Global South, 51–52; role of donors in, 52; in suburbs of the Global North, 52

Returning Farmland to Forest Program (RFFP, Grain for Green Program, China), 102–6

roads, 20, 42; abandonment of, 70–71, 102; and deforestation, 101–2; logging, 101–2; and forest plantations, 99

Roosevelt, Franklin Delano, 37. 80

Roundtable on Sustainable Palm Oil (RSPO), 113; effects of, 113. *See* ecocertification

rubber: and monoclonal rubber plantations, 155; and sisipan,153; in Southeast Asia, 151–57. *See* shifting cultivation

Sangay National Park, Ecuador, 43–45

Schmitter, Philippe, 25. *See also* corporatism

scientists, in social movements to conserve rain forests, 49–50, 202

secondary forests: and mountainous terrain, 91–92; and rural to urban migration, 68–96

settlers, 15–17, 52; from the Andes, 42–45; in New England, 69–73

shifting cultivation, 151; and carbon sequestration, 39; and jungle rubber, 151–53

Shuar, xiii, 8; federation of Shuar villages, 45; household economies, 46; resistance to extraction and development interests, 46; role of in creating protected areas, 42–48; trees in Shuar pastures, 135–36

silvopastures, 19, 128–48; in Babassu, 137–41; carbon sequestration in, 136; project in Colombia, 136–37; rain and tree seedlings germinating in, 131; on rented land, 135; trends in areas occupied by, 129;

social movements, 201–5; and activists, 201–5; to conserve tropical rain forests, 42–48; institutionalizing for forests, 205–12; to stabilize the climate, 201–5

societal corporatism, 22–29, 94, 127; in business cycles, 25; in post-World War II Scandinavian countries, 25

Socio-Bosque, jurisdictional approaches to REDD+, 47. *See* payments for environmental services (PES)

soil exhaustion and erosion, in deforested areas, 77–82

South America, 12; cattle ranching in, 130; pastoral landscapes in from the colonial era, 130

Southeast Asia, 12; and carbon sequestration, 4, 39; and fires, 2; and oil palm production, 110–14; and rubber production, 152–55

state corporatism, 25, 127; in China, 25, 102–6; and fascist regimes, 25. *See also* Schmitter, Philippe; societal corporatism
state strength, 120–21; and planted forests, 121; weak states and Great Green Wall in the Sahel, 146
sustainable agricultural development, 171–74
Sustainable Cocoa Production Programme (SCPP, Indonesia), 163

Tanzania, 29; compact with Norway, 29; and local autonomy in governance, 29
Thunberg, Greta, 21, 201. *See also* countermovements; social movements
timber companies: in Chile, 118–19; in Indonesia, 112–14; in Laos, 109–10; in United States, 114–15
top-down forest initiatives, 5–7
tree tenure, 144

United Nations: Ecosystem Restoration Decade, 5. *See* Framework Convention on Climate Change; Convention on Biodiversity
United States: abandoned cotton fields in, 77–82; Civilian Conservation Corps, 81; Conservation Reserve Program, 93, 116; and forest fires, 3; National Forests, 81; northeastern land abandonment in, 69–73

vapor pressure, 2
Vietnam: and Doi Moy reforms, 106–7; forest plantations in, 106–9; and the National Five Million Hectare Reforestation Program, 108; and state forest enterprises, 106; the 327 Program, 107

woodlands, open and semi-arid, 141–46
World Bank, 9, 207
world society theory, 21, 204
World War I, and corporatism, 25

Yangtze River floods (1998), 102–3